Land Degradation and Desertification

Land Degradation and Desertification

Edited by

V.C. Jha

RAWAT PUBLICATIONS
Jaipur and New Delhi

ISBN 81-7033-825-5

Published by
Prem Rawat for *Rawat Publications*
Satyam Apts., Sector 3, Jawahar Nagar, Jaipur 302 004 (India)
Phone: 0141 265 1748 / 7006 Fax: 0141 265 1748
E-mail: info@rawatbooks.com
Website: www.rawatbooks.com

Delhi Office
4858/24, Ansari Road, Daryaganj, New Delhi 110 002
Phone: 011-23263290

Typeset by Rawat Computers, Jaipur
Printed at Chaman Enterprises, New Delhi

Contents

Preface

The International Geographical Union (IGU) formed a
Commission on Land Degradation and Desertification in 1996.
The main aim of the IGU was to achieve a better practical and
theoretical understanding of land degradation and desertification
processes that affect the world's landscape. In fact, increased food
production in tropical countries is essential to feed the enormous
population. Intensification of agriculture has resulted in drastic
changes in landuse patterns affecting denudational processes. In
recent years, land degradation and desertification have become
common problems due to climatic conditions and changes in
landuse, particularly in the tropical lands. It is noted that every
year, millions of hectares of land suffers from physical, chemical
and biological degradation which ultimately reduce its current and
potential capability of goods and services through agriculture,
engineering, sanitary and recreational uses. The term 'land
degradation' relates to the loss of utility or potential utility or the
reduction or the changes of features or organisms or compositions
which cannot be replaced. In fact, land degradation is responsible
for reducing the status of the materials which are the result of
various natural and anthropogenic factors. The process of land
degradation not only reduces the capability but also accelerates the
process of land 'desertification'. The term 'desertification' is being
used to explain the changes which take place over the surface due

to natural and man-induced processes. This ultimately results in rendering the land either infertile or not at all suitable for agriculture, and to some extent for human habitat as well.

At the global scale, the process of land degradation is visualised from soils, water, natural vegetation, wildlife, ecology and human habitation points of view. The physical, chemical and biological degradation of land relate to all the above factors.

Various studies have been attempted to study land degradation with a view to have sustainable land management. In fact, land degradation is a central challenge to sustainable development. The latter has been defined as development that "meets the needs of the present without compromising the ability of future generations to meet their own needs." This was accepted as a common goal at UN Conference on Environment and Development in 1991. At large scale, the main problems threatening natural resources and the sustainability of life support system are soil degradation, availability of water and the loss of biodiversity.

The problem of land degradation and desertification is being studied widely and the conservation and rehabilitation of the resultant land are being taken care with scientific measures and technology. The role of spatial information technology is very vital in gathering the land degradation and desertification data. The cross fertilisation of remote sensing, GIS and land degradation and desertification will provide a sound environmental assessment, paving the way for sustainable land management in the years to come.

I am very grateful for the cooperation of the participants who presented their paper in the Twenty-Second Conference of the Institute of Indian Geographers (IIG) and International Geographical Union (IGU) Commission meeting on Land Degradation and Desertification (COMLAND) organised by me during January 9-11, 2001 at the Department of Geography, Visva-Bharati University, Santiniketan, West Bengal, India. Among the presented papers, 26 selected papers are included in this volume and I thank the contributors who very carefully prepared the manuscripts in time. This IIG Conference-cum-IGU COMLAND Meeting was made possible with the financial assistance and sponsorship from a number of reputed research institutes and government departments of the country such as Council of Scientific and Industrial Research, New Delhi; Department of Science and Technology & NES, Kolkata; Department of Land Resources, Ministry of Rural Development, New Delhi; Indian Space Research Organization, Bangalore; Anthropological Survey of India, Kolkata; Development and Planning Department, Government of West Bengal, Kolkata; National Institute of Hydrology, Roorkee: National Remote Sensing Agency, Hyderabad; PCI, Kolkata and National Atlas and Thematic Mapping Organisation, Kolkata. I thank them all. I take this opportunity to thank Rawat Publications for publishing this volume in a very short time. My thanks are also due to Prof. P. Jash, Prof. K.C. Gupta, Prof. A. Neogy, Prof. G. Subbiah, Prof. A. Biswas, Prof. S.J. Stephen, Dr. S.N. Ojha, Mr. S.K. Dutta, Mr. P. Roy, Dr. U.S. Malik, Dr. Saswati Kapat and Dr. Bhaswati Mukherjee for their help in various ways.

V.C. Jha
Full Member
IGU Commission on Land Degradation & Desertification

Introduction

V.C. Jha

This volume is an outcome of the Twenty-Second Conference of the Institute of Indian Geographers (IIG) and International Geographical Union (IGU) Commission Meeting on Land Degradation and Desertification (COMLAND) held at the Department of Geography, Visva-Bharati University, Santiniketan, West Bengal, India during January 9-11, 2001.

In the first paper, Maria Sala has studied in detail the surface hydrology, sediment transport, and water quality of representative landuse units at a catchment level. She has also discussed the point-source pollution related to agricultural, industrial, urban and tourist activities. The main focus of this paper is on the environmental implications for water management and storage which are very important for development and planning of a region.

Dagmar Haase, Annegret Haase, Oliver Spott, Lothar Linde and Thilo Weichel have tried to assess the current state and forms of ecological as well as social degradation in the arid Andean regions of north-western Argentina. It focuses on the historical development of both farming and pastoral agricultural system in the valley of Huma-huaca and the Argentina Pima. The current change in landuse during ENSO events, a highly variable rainfall intensity, problems of land ownership and socio-economic status of the indigenous Kolla population, have been explained with the help of maps and tables. Lastly, it focuses on the problems and prospects of the future development of these rural, marginal and degraded mountain areas.

R.Y. Singh has attempted to study the causes and consequences of land degradation in arid, semi-arid and dry sub-humid areas due to climatic variations and human activities. He has made a statistical assessment of desertification around the world with the help of data provided by UNEP as well as United Nations' Plan of Action to combat desertification. He has also dealt with the various financial constraints regarding the implementation of the recommendations and has focused on the need of future steps to remove them.

N. Batnasan has made a study on land degradation in the Valley of Lakes of Gobi desert region in Mongolia. He has presented the seasonal and long-term changes of river water resources and lake levels, irrigation activities and related degradation problems. The morphological characteristics of lakes, seasonal variations of precipitation, river hydrology characteristics and irrigation activities have been shown with the help of maps and diagrams. Finally, he has suggested some measures for the careful management and conservation of water resources in this region.

N.K. De and Ananya Taraphder have given a conceptual overview of physical and chemical degradation of land by natural as well as human agents and the resulting problems. The factors and types of land degradation have been classified in different categories with particular reference to technology, which though improves the quality of life but lowers down environmental conditions in many cases.

The changing landuse/landcover scenario in Nepal has been studied by Motilal Ghimire and Narendra Raj Khanal. Information on landuse cover in the country reported by various authors from different parts of the country in different times has been collected, compiled and analysed. The causes, processes and the rate of landcover/landuse change in the country have been synthesised with the help of charts and tables.

The tremendous population expansion has resulted in the deterioration of surface water bodies as well as ground water resources. In his study, D. Das has made an attempt to appraise the landforms and locate the sites of ground water development using remote sensing and GIS techniques. The author has discussed how the water table conditions can be improved in the hard rock region and surface storage be developed through artificial recharge and water harvesting. In this context, the drought-prone river basins of Kangashabati in Purulia and Sali in Bankura have been studied in detail. Composite maps have been generated from visual interpretation of satellite data. This study is helpful in preparing a better water management strategy.

The biological influences in geomorphological processes have resulted in producing special soil features and regulating the soil-forming mechanisms and destroying soil fabric, mixing soil horizons and creating micro-relief by mounding and excavation. The termites play a very important role in the creation of micro-relief and their rate has been assessed by R.G. Patil who has made a case study of mounds around Pune. Termite mounds are the most striking manifestation of termite activity and are enduring structures. He has tried to assess the role of geographical factors on mound, soil characteristics between mounds and surrounding areas and distribution of mounds with chain and tape survey. He has also made a mechanical and chemical analysis of mounds and surrounding soils. The different species of termites, length, height, north-south construction of mound, pH values, etc. have been discussed in detail with the help of tables and graphs. His study will be of great help to evaluate the influence of plants and animals on landforms.

Accelerated gully erosion and other forms of water induced soil erosion are constantly converting usable land into waste-land in India. The work by V. C. Jha and Saswati Kapat determined the nature, extent and distribution of gully network in association with climatic pedo-genesis, geomorphic and anthropogenic aspects in the Dumka district of Jharkhand with the help of satellite imageries, toposheets and intensive field work. The formation of

gully scars are particularly noticed in south-west monsoon season
when it receives the maximum amount of rainfall with high
intensity, in this area composed of gneissic rocks. This area falls
under moderate morphogenetic region. The assessment of profiles
in gullies, linear and side wall erosion of gullies, analyses of soil
samples have been made, volume of eroded materials from gully
walls and the major gully affected areas have been shown with the
help of maps and tables. Lastly, they have suggested some anti-
erosion measures like damming of gullies, plantation of trees along
the sides of gullies, etc. It is very essential to properly utilise the
wastelands and the restoration, conservation of village resources is
necessary to attain sustainable environmental development.

Kuntala Lahiri-Dutt and Prasanta K. Jana have given a report
on how mining causes land degradation in the coal mining region
of Raniganj. Active mining that has resulted in removal of forest
cover, soil erosion and water and air pollution, decline in land
potentiality and agricultural production, have been discussed in
detail. The decline in potentiality of land has being assessed with
decreasing land-mass ratio and increasing population. To meet the
increased demand for coal, private entrepreneurs intensified
mining operations in an unplanned way, and even nationalisation
of collieries could not check degradation. The region is dotted with
abandoned collieries, and these underground voids of unknown
depth often lead to land subsidence and mine fissures.

A similar study of land degradation due to mining activity in
Raniganj coalfield has been made by Debasish Sarkar. He has
analysed in detail the physical and anthropogenic causes of
degradation in this region and has provided an year-wise estimation
of land degradation. He has focused on the systematic coordination
between pre-operational and post-operational phases of mining,
proper balance between extraction of coal and measures to control
land degradation to attain sustainable development in the region.

As already mentioned, anthropogenic factor is an important
factor for land degradation. In this context N. K. De and Ananya
Taraphder have made a case study of land degradation in Burdwan
with particular reference to human activity. The district of

Burdwan has a tropical environment and almost every year recurrent flood and drought cause negative impact on the economy of the district. Most of the rivers are influenced by human actions and this is the dominating feature of the area. Their study reveals that degradation is maximum in the central rolling plain due to mining activities, relatively steep gradient; deforestation has caused degradation in the west; and in the east, agriculture, urbanisation and river behaviour are responsible for degradation. Human influence is strongly felt through an account of transformation of vegetation from native to cultural ones. Types of land degradation by human activities have been classified into four main types, namely, mining which causes land subsidence, loss of agricultural land, industrial air and water pollution; agricultural practices which deteriorate the physical and chemical properties of soil and contaminating surface and municipal ground water creating exhausted sub-surface water layers resulting in subsidence and slow process of humificaton etc. Lastly, the authors have suggested certain strategies to overcome land degradation in Burdwan, laying stress particularly on the need and considered necessity to control human activities not only by enacting laws but also by creating awareness among common people.

Among the various types of physical parameters, slope plays an important role in determining the thickness of soil, its fertility, texture, permeability, drainage condition, vegetation coverage and agricultural landuse. It is a dominating factor in areas which are undulating and dissected having a high degree of slope gradient.

A study of the Damodar-Dhajmma interfluve of Burdwan and Purulia districts of West Bengal area is attempted by Nageshwar Prasad and Manjari Sarkar (Basu). They have drawn correlation between the characteristics of slope and aspects of agricultural landuse. Climatically, the region falls under tropical climatic condition and rivers are mainly controlled by original slope. The authors have classified six major slope categories ranging from 0.5-2.5 and correlated them with five landuse pattern categories by means of bi-variate analysis based on field observation and block-wise statistics. Maximum amount of cultivated area lies in the 0.5 –1.5 slope zone. Crisis of employment has resulted in a shift to agriculture causing overpopulation and economic inequality in the gentle and moderate slope zone. The cultivated lands decrease with increasing slope but now agricultural land is also decreasing to settle the increasing population.

Onkar Prasad has made a study of how soil structures have promoted land degradation in the Hazaribagh plateau. He has studied in detail the various types of soils, their distribution, fertility and productivity with the help of maps. This region faces high degree of soil erosion because of highly uneven topography, intense rain, excessive run-off, low soil permeability, deforestation and overgrazing. The worst affected areas are the Giridih uplands where degradation is largely due to gully erosion. Only the steeper scarp lands with vegetal cover have remained intact due to inaccessibility to man and livestock. He has provided certain soil conservation proposals adopted by DVC and state government to check soil erosion.

In the next paper, S. Gangopadhyay makes a critical study of the management procedure resorted to by the local people of North Sikkim in order to make the degraded land resourceful. The author analyses in detail the factors of degradation in this region and the traditional techniques by means of which the local people are trying to combat the crisis. Here degradation is mainly due to steep slope, landslides, soil erosion, prolonged winter, waterlogged condition, etc. The people, to check degradation, use a variety of traditional techniques of sowing seeds, cutting trees at certain height to prevent soil erosion, laying of fallen trees along the contour lines to regulate water, etc. This is a completely different kind of study which reflects how often old and traditional methods become very useful in checking excessive land degradation.

Premature reclamation has resulted in several socio-economic problems in the world's largest delta in the Sundarban. In this context, S.N. Chatterjee has made a detailed analysis of the evolution of the physical environment of the Sundarban, the various phases of reclamation, and the ill-consequences of primitive reclamation that has increased flood effects and salinity. The mangrove forests of Sundarbans have great economic value and therefore it is necessary to preserve them with the help of appropriate measures. For this purpose, the author stresses on the involvement of people at the grass root level to save their own lives.

The next paper, by Susmita Ghosh and Subrata Ghosh, is based on their experiences of land degradation due to excavation of lateritic terrain for 'murrum' and gravel for the road construction and industrial purposes near Durgapur. This has resulted in the removal of original vegetation cover and damage to young plants, ultimately causing desertification. The entire area looks like badlands densely covered with pits and groves of varying sizes. However, there are various ways in which these gravel and the land can be restored. The authors have highlighted some of them that should be immediately taken into account by the concerned authorities to prevent this kind of desertification.

V.C. Jha and Saswati Kapat have also tried to present an estimate of the soil loss on the basis of the universal soil-loss equation, particularly red soil and lateritic terrain in Dumka area. It has also helped in assessing the degree of land degradation in the study area. Similarly, in another study V. C. Jha and K. Gupta have explained the causes of land degradation in the Dwarkeshwar basin as an example of the tropical lands.

Gear M. Kajoba has mainly focused on the land and natural resources tenure reforms in developing countries. This paper highlights not only the historical aspects but also the facets of agro-forestry. He has also presented comparative facts in selected African countries. Similar attempts have been made by Mirza Mafizuddin and Mesbah-us-Saleheen in their paper focusing on the status and the problems of environmental degradation in the Madhupur tract of Bangladesh.

An attempt to estimate one-day maximum rainfall for 2 to 100 years return periods and one-day probable maximum precipitation using all long period stations in West Bengal has been made by B.N. Mandal, R.B. Sangam and A.K. Kulkarni.

Avijit Gupta has discussed the exploration linkage in the tropics and sub-tropics with land degradation, environmental impacts and fluvial geomorphological aspects.

Large number of dams, barrages and embankments are constructed along the rivers to protect the agricultural land from floods as well as for supplying irrigation water, but often they become important causes for river-bank erosion and degradation, as indicated by the experiences of Sutapa Mukhopadhyay along the river Ganga near Farakka barrage. Steep stream gradient and high velocity of the river due to the presence of hard granitic rocks of the Rajmahal, and breaching of embankments are important causes of rapid bank erosion, but statistical data show that after the construction of the Farakka barrage land erosion accentuated largely causing devastating floods in the Maldah and Murshedabad districts of West Bengal. She has also considered the increasing pressure of population on land as an important cause for this environmental crisis. Lastly, she has suggested certain measures to overcome this crisis and proclaimed that people should try to adjust with the dynamic behaviour of the river.

A similar study of floods and land degradation has been made by Mallanath Mukherjee and Malay Mukhopadhyay based on their experiences in the lower Ajoy river basin. Flood has become a regular feature of this basin. The most devastating floods occurred in 1956, 1978, 1995, 1999 and 2000 causing tremendous physico-economic losses. They have studied in detail the physical and anthropogenic causes of floods and soil erosion in this area. Flood levels at different gauge stations, spatial distribution of sand splay, soil status report, loss of cultivated land have also been assessed and the results are presented with the help of tables and maps. The authors have explained in detail the causes, intensity and effects of degradation and suggested some useful measures to mitigate the problems.

A unique attempt has been made by Mahesh K. Gaur and Hemlata Gaur in which they have discussed the anthropic activities impetus in creating the land degradation and desertification.

1

Landuse Impacts on Water Quality in a Mediterranean Drainage Basin

Maria Sala

The Mediterranean environment, due to its climate and relief characteristics, it is easily degraded both by natural and by anthropic actions. The result is the alteration of the hydrological, biological and pedological systems and, hence, an activation of desertification processes. Changes in landuse have taken place along history in relation to socio-economic changes. At present its importance and its impact on the environment are extremely outstanding. On the one side, an abandonment of silviculture is taking place, for instance, the extraction of the bark from the cork trees, and the abandonment of the traditional agriculture which complemented silviculture. On the other side, the mountain is more and more occupied by uses related directly or indirectly to tourism, for instance house and road building, and development of recreational resources. In the lowlands intensive agriculture and breeding are progressively taking place.

Progressive urbanisation of forested areas and small villages, and the changes in the agro-forestry exploitation have consequently a change in the hydrological response of the land surfaces because they alter the relationships between rainfall, infiltration, runoff, and sediment transport. The traditional landuses have given way to an accelerated process of urbanisation which implies an increase of impervious surfaces, hence of runoff. Later, an increase of surface runoff produces and increase of erosion on slopes and fields and, as a consequence, a sedimentation process on the stream channel and an increment of the risk of flooding in the lowlands.

Objectives

The aim of the research is to study the surface hydrology, sediment transport, and water quality of representative landuse units at a catchment level. Particular attention is paid to point source pollution related to agricultural, industrial, urban, and tourist activities, and its environmental implications for water management and storage.

Study Area

The study area is located at the north end of the Catalan Coastal Ranges and the Selva depression, east of the city of Girona. Geographical coordinates are: 41°54'-41°46'N and 3°4'-2°50'E (Figure 1). The Ridaura basin is here considered representative of the different landuses in the area, such as forest, agriculture and touristic activities.

Physical Trends

The drainage basin of the Ridaura river has an area of 73.8 km^2 (Figure 2). The highest points in the basin are between 400 and 500m asl. The length of the main stream draining the catchment area is 20 km. Mean slope of the streamchannel at the headwaters is 5% decreasing to 2% at the bottom of the Aro valley, and to almost 0% at the outlet. It comprises the northeastern slopes of the Cadiretes massif and the southern slopes of the Gavarres massif. These two massifs surround the tectonic depression of the Aro valley, and they constitute the northbound of the Catalan Coastal Ranges. Structural fracturation of the area follows NNW-SSE and NW-SE direction. The main stream flows from the Cadiretes massif down to the centre of the depression and to the Mediterranean sea.

Figure 1
Location of Study Area

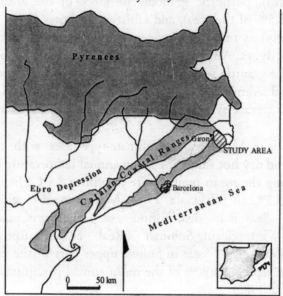

Figure 2
Ridaura Drainage Basin

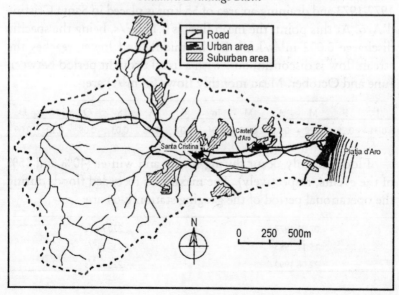

Lithology is mainly granitic, composed by leucogranite (41.5% of the area), porphiric granodiorite (26% of the area), biotitic granite (8.5% of the area), and schist and arkoses (2.4% and 0.3% respectively) as minor components. In addition, there is a large variety of dykes. Alluvial deposits occupy 18.2% of the basin area. Due to the granitic rock and the low energy of the slope most of the alluvial sediments are sands. In its middle reach the Ridaura has built up a terrace system which is linked to the slopes by alluvial fans.

This is a Mediterranean climate-type area with temperate winters and dry hot summers. Mean annual temperature is around 16°C, being the mean temperature in winter 8.8°C (January) and almost 23.3°C in July (Sala, 1979). Mean annual precipitation is 616mm. Seasonal distribution of the precipitation is Autumn-Winter-Spring-Summer. Real evapotranspiration is around 660 mm per year in Solius, upper part of the catchment area. This represents 90% of the mean annual precipitation in the area (Trilla, 1980)

Information about hydrology of the Ridaura is supplied by the Water Authorities. A gauging station (E-64), in service since 1972-1973 and draining an area of 55 km^2 is placed in Santa Cristina d'Aro. At this point, the mean flow is 0.13 m^3/s, being the specific discharge 0.002 m^3/s/km^2. In the middle and lower reaches the streamflow is discontinuous, with a usual drought period between June and October. Mean monthly flows (in m^3/s) are:

J	F	M	A	M	Jn	Jl	A	S	O	N	D
0.31	0.28	0.22	0.21	0.17	0.04	0.01	0.00	0.03	0.08	0.19	0.09

Floods mainly occur during spring and winter (40% and 35% of the events, respectively). The maximum recorded floods during the operational period of the gauging station E-64 are:

07.01.1977	220 m^3/s
15.11.1983	235 m^3/s
22.06.1992	225 m^3/s

The aquifer consists of a mantle of alluvial sediments lying on the bottom of the Aro valley. They have a volume of 160 Hm^3, being 130 Hm^3 of them saturated. Conventionally, the aquifer is divided in two main parts: the reservoir of Castell d'Aro and the reservoir of Santa Cristina d'Aro (Ministerio de Obras Publicas, 1969). Average hydraulic conductivity varies from 48 m^2/day and 1330 m^2/day (Trilla, 1980).

The water consumption is about 5 Hm^3 per year, 70% for domestic uses, and 30% for irrigation. Water demand rises up during the summer period. At present, and due to the increasing demand of water and its low quality for human consumption, water from the Ter basin (El Pasteral dam) is diverted to the Ridaura river basin. Average discharge of the transfer is 0.05 m^3/s.

Vegetation it mainly consists of a cork-oak *(Quercus suber)* forest with evergreen-oaks *(Quercus ilex)* and pines *(Pinus pinaster, Pinus pinea, Pinus halepensis)*. At the headwaters the vegetation is very dense. There are plantations of poplars and eucalyptus along the stream network.

Human Activities

Until the beginning of the 20th century the main economic activities were forest (cork tree, coal) and agricultural exploitation. With the onset of tourism the economic activity shifted to the tertiary sector. At present forests occupy 61% of the catchment area, followed by agricultural fields (21%), and other activities (17%) mainly urban areas, related tourist resources (golf courses and urbanisations), and paved and non-paved mountain roads.

The population is mainly concentrated at the bottom of the Aro valley in the villages of Santa Cristina, Castell d'Aro and Platja d'Aro. The municipality of Llagostera is part of the Ridaura basin, although the village of Llagostera is out of the catchment area. In the mountain areas population is dispersed. The evolution of the resident population along the 20th century is as follows:

	1900	1930	1950	1970	1981	1986	1990
Santa Cristin d'Aro	1069	1034	985	980	1281	1446	1483
Castell-Plantja d'Aro	1185	1128	1128	2473	3778	4243	4676
Llagostera	4140	4090	3812	4464	5033	5119	5124

The increment in the municipality of Santa Cristina in the centre of the basin during the century is almost 40%, and the increment in the municipality of Castell-Platja d'Aro in the coastline is about 260%. In this municipality the increment becomes spectacular from the seventies onwards. Since 1970 the increment is of 110 new residents per year.

If we take into account the number of the non-resident inhabitants, it is the tourism which comes to the area mainly during summer months, the increment of population in these municipalities becomes even more relevant. During summer months, the population in the municipality of Santa Cristina d'Aro multiplies by 7.5, while in the municipality of Castell-Platja d'Aro the increment is by 20.

Methodology

Field data has been obtained weekly at vigil sites and during flood events. The sampling points to control discharge, sediment transport and water quality have been selected attending to landuse and point source pollution. For landuse impacts three sectors were selected representing forest, agriculture and urban landuses. In addition, six sampling points were selected in order to monitor point-source of pollution.

Weekly measurements of the streamflow were taken in the sampling points along the stream. The velocity has been measured by means of an automatic OTT C-2 current meter. At low flows volumetric measurements were performed. In the gauging station E-64 continuous records of water depth are available.

Manual sampling of water and sediment was also performed once a week. In the field temperature of the water, pH, electric conductivity, alkalinity, and dissolved oxygen (DO) were measured with the following instruments: temperature by a digital thermometer, pH by digital pH meter Hanna, conductivity by digital condutimeter Hanna (HI 9033), dissolved oxygen by digital oximeter Dinko (Oxan), alkalinity by Kit Merck and at the laboratory by titration. Finally, suspended sediment samples were taken by means of a U.S. DH48 Depth-Integrating Sampler.

Chemical analyses of water were made following the standard methods reported by SUESS (1982) and Environmental Monitoring and Support Laboratory-EPA (1983). 1000 ml of water are filtered with a 0.45 µm milipore cellulose ester. Part of the sample (for cations determination) was stored with a solution of 1% HNO_3, and part of it (for silica and anions determination) was stored without adding any solution. All the subsamples were kept into a cleaned standard containers and placed in the refrigerator (4°C), ready to be analysed. Silica determinations were usually made either 24 hours or 48 hours after the sample collection and never later than one week. Cations and anions determinations were made, depending on the equipment's timetable, no later than 15 days after the sample collection.

Techniques used to determine chemical components of the water and their concentration are: Inductive Coupled Plasma (ICP) for major cations (Ca^{++}, Na^+ and Mg^{++}) as well as for heavy metals (As, Hg, Cd, Pb, Mn, etc.). Major anions have been analysed by ICP technique. Potassium was determined by Atomic Absorption Spectrography. The quality control of chemical analysis has been tested by means of an ionic balance. Samples with a variation between Scations/Sanions higher than 15%, due to failures either on samples preparation or on analysis procedure, have been rejected.

The sediment was deposited after filtering the water in a Milipore 0.45 µm mixed cellulose esters. The amount of deposited sediment was calculated weighting the filter by means of an electronic balance. Concentrations of suspended sediment are given in mg/l.

Results

Hydrological Characterisation of the Drainage Basin

The hydrological characterisation of the Ridaura drainage basin has been produced on the basis of the rainfall and runoff data of the period 1972-1986 obtained at the gauging station E-64 located in Santa Cristina (Sala et al, 1994).

(a) Annual Rainfall-Runoff Analysis

	Rainfall	Runoff	Losses	Runoff Coefficient
1972-73	501.4	79.8	421.6	15.9
1973-74	544.4	91.6	452.7	16.8
1974-75	731.2	59.6	671.6	8.2
1975-76	581.2	32.2	549.0	5.5
1976-77	998.4	273.6	724.8	27.4
1977-78	660.5	66.2	594.3	10.0
1978-79	526.3	82.5	443.8	15.7
1979-80	632.7	84.7	548.0	13.4
1980-81	565.3	34.7	530.6	6.1
1981-82	557.8	79.6	478.2	14.3
1982-83	485.5	23.4	462.0	4.8
1983-84	732.4	79.4	652.9	10.8
1984-85	381.2	10.3	370.8	2.7
1985-86	733.0	109.6	623.4	14.9
1986-87	612.7	66.7	546.0	10.9
Mean (mm)	616.3	78.2	537.9	12.6
STD (mm)	145.1	60.7	101.3	6.2
Coeff.Var. (%)	23.5	77.6	18.8	52.4

The coefficient of variation indicates an important scatter of points around the mean, especially for the runoff observations. The mean annual runoff represents only 12.6% of the mean annual precipitation. This fact indicates that the evapotranspiration in this area is very important. The relatively low correlation between rainfall and runoff can be due to the high interannual hydrological variability in the drainage basin and, perhaps, to the small number of available data.

From these data, and taking into account an average infiltration rate along the catchment area of about 1 Hm^3 per year, an approximate water balance can be proposed:

	(mm)	(%)
Annual rainfall	616	100.0
Annual runoff	78	12.6
Annual losses	538	87.4
ETR (*)	520	96.6
Infiltration	18	3.4

(*) Real evapotranspiration below the data reported by Trilla (1980)

(b) Analysis of Monthly Discharges

Average monthly precipitation (P) and runoff (R) data (in mm) for the hydrological period 1972-1986 are the following:

	O	N	D	J	F	M	A	M	Jn	Jl	A	S
P	73.9	58.7	32.1	53.4	47.8	63.5	45.6	67.5	35.3	29.6	47.5	61.2
R	3.8	8.7	4.1	14.7	13.1	10.1	10.0	7.9	1.7	0.3	0.1	1.5
%	5.1	14.8	12.6	27.5	27.4	15.9	22.1	11.7	4.7	0.9	0.3	2.5

From these observations, some considerations about the monthly distribution of rainfall and runoff can be made:

(a) Except for January, February and April, the mean monthly runoff represents less than 20% of the monthly rainfall

(b) Runoff/Rainfall ratio is extremely low during summer months, corresponding to high temperatures and, thus, to an intensive evapotranspiration in the area

(c) December and May show a similar runoff coefficient than the annual mean runoff coefficient

Downstream Changes in Runoff

A homogeneous series of 10 discharge values obtained from January to December 1993, mainly during base-flow conditions, is

Figure 3
Sampling Points in Relation to Landuse

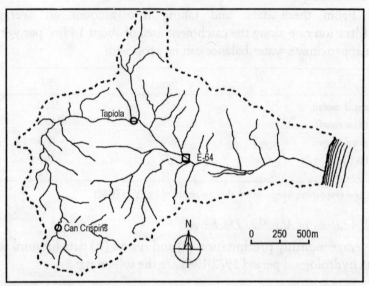

presented. The drought period lasted between May and November 1993.

Landuse	C. Crispins forest	Tapiola forest + agriculture	E-64 Urban
Area (km^2)	2.8	25.5	55.0
Average discharge (m^3/s)	0.072	0.639	0.947
Specific discharge (m^3/s/km^2)	0.026	0.025	0.017

A downstream reduction of the specific discharge is observed, due to high infiltration capacity of the lowland areas compared to the upstream areas. Besides, an increase of the water demand, especially for domestic and tourist consumption has to be considered in order to explain the reduction of the specific discharge downstream.

Sediment Transport

Average suspended sediment concentration (mg/l) and variability at the vigil sites obtained weekly every during 1993 and 1994, are:

	Mean	*Standard Dev.*	*Coeff. Var. (%)*	*N° Samples*
C. Crispins	5.2	2.7	52	16
Tapiola	81.7	113.9	139	16
E-64	90.1	127.4	141	38

There is a downstream increment of the suspended sediment concentration. It is lower at the forested upper part of the drainage basin (5 mg/l), where the influence of the human activities is low, while it markedly increases downstream once the stream flows in the middle (82 mg/l) and lowlands (90 mg/l), where the human activities (agriculture, road works, urbanisation, non-paved roads) become more intensive. The variability of the suspended sediment concentrations is very high, of the order of 100%.

The sediment yield of each of the monitored sub-basins has been calculated by multiplying the mean discharge times the mean suspended sediment concentration. Results for each of the vigil sites along the stream are as follows:

	tonnes/ha/yr
C. Crispins	0.05
Tapiola	0.55
E-64	0.46

Downstream increment of the specific sediment yield associated with landuse changes is similar than in the Verneda (Sala et al., 2000), although differs from other drainage basins, like Arbucies and Tordera, which also include a lowland area with agricultural activities (Batalla and Sala 1996). In those rivers, the sediment yield decreases as the drainage basin increases, mainly due to downstream sedimentation and to the lower capacity of transport of the streamflow. Therefore, human activities, mainly agriculture and urban (the construction of a new road surrounding the village of Santa Cristina d'Aro is a good example), in the middle and lowland area of the Ridaura basin changes the general pattern of sediment yield observed in most drainage basins.

Dissolved Load at Base Flow Conditions

Average results of the dissolved load, obtained every week during 1993 and 1994 are described in this section. Data has been collected at the vigil sites of Can Crispins (representing forest landuse), Tapiola (representing forest and traditional agriculture), and E-64 at Santa Cristina d'Aro (within a partially urbanised area).

In the upper reach (Can Crispins) the water chemistry shows a mean concentration of 79.9 mg/l dominated by chlorite (25%), followed by carbonates (18%), sodium (17%), sulphate and silica (14%), calcium (6%), magnesium (2.2%), potassium (1.2%), and nitrates (1%) as minor components. From these data two remarks can be made. The high concentration of chlorite and sodium can be related to the proximity of the Mediterranean Sea (3 km).

In the middle reach (Tapioles), the mean concentration of dissolved load increases to 184 mg/l, dominated by carbonates (38%), followed by chlorite (14%), calcium (12%), sulphate and

Figure 4

Downstream Changes of Mean Concentration of Solutes and Suspended Sediment at Base Flow Conditions (1993 and 1994)

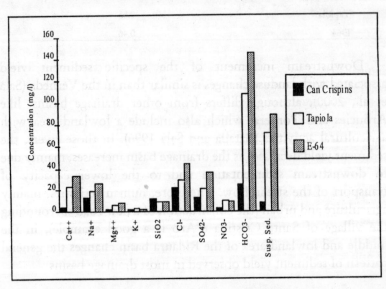

sodium (10%). Nitrate, silica, magnesium and potassium (1.4%) are minor components.

In the lower reach (E-64), concentrations are dominated by carbonates (40%), followed by chlorite (18%), calcium (17%), sulphate (14%), sodium (10.5%). Nitrate, silica, magnesium and potassium are minor component.

Summarising this, during base flow conditions, water chemistry changes in the Ridaura basin in the sense that there is a progressive downstream increment of the concentration of calcium, sodium, and magnesium. Chlorite and sulphate concentrations increase, due to the downstream increment of domestic inputs (urban waste and sewers from the towns). The concentration of chlorite increases downstream, 28 mg/l being the recommended value for potable drinking water. Moreover, the concentration of sulphate increases downstream, 25 mg/l being the recommended value for drinking water.

Nitrate increases between the forested and the agricultural area, but decreases downstream once the river enters the urban reach of the drainage basin. Similarly, the mean electrical conductivity at 25°C increases from 176 µS/cm in the forested area (Can Crispins), to 343 µS/cm at the vigil site of Tapiola, and to 463 µS/cm at the vigil site of E-64. Nitrite concentrations at E-64 indicate recent pollution. Phosphates have not been found. No significant values (< 1 mg/l) of heavy metals (Fe, Al, Mn, Zn) have been found.

Downstream Changes in Water Quality

This section presents the results on water quality obtained between February 1994 and June 1994 in the Ridaura river basin (Martin, 1994). For this purpose, we have considered the European Board 80/778/CEE (B.O.E. 20th September, 1990), which determine the pattern and the acceptable limits to ensure water quality for human usages. Next, a summary of the obtained results from fifteen samples:

Figure 5
Sampling Points in Relation to Water Quality

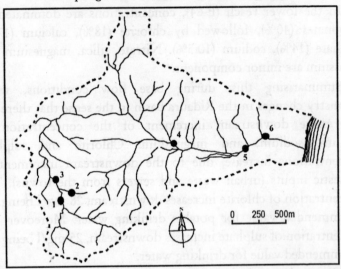

(a) Point 1: It is located at the forested headwaters of the catchment area, and it does not receive any meaningful anthropic interference. However, it can occasionally suffer some interferences from the Solius rubbish-dump.

(b) Point 2: This point is located at the outlet of the stream which drains the Solius rubbish-dump. All chemical parameters are widely disturbed:

— Electric conductivity: 100% of the samples showed levels higher than the suggested values (400 μm/cm). Seven samples had conductivity values higher than 800 μs/cm, and two samples had conductivities of 1080 μs/cm and 1258 μs/cm.

— Dissolved oxygen: Results are within the accepted limits. This fact can be explained by the turbulence of the stream. Only one sample presented a concentration of 2.6 mg/l, which is lower than the limit (3 mg/l).

— Biological Oxygen Demand (BOD): 100% of the samples presented a remarkable consumption of oxygen. For all samples, concentration of oxygen five days after the sampling day was lower than 3 ppm, and eleven samples had values ranging between 0.4 and 0.6 mg/l.

— Chlorite: All samples presented results higher than the level guide (25 mg/l). Concentrations varied between 111,30 mg/l and 383,6 mg/l.

— Ammonium (NH4): All samples showed values higher than the maximum permitted *concentration* 0.5 ppm. Results were between 1.3 and 11.1 mg/l.

(c) Point 3: This sampling point is located downstream the junction of the Ridaura and the tributary from the Solius-dump. Only a small influence of the rubbish-dump can be observed and attributed to a dilution effect.

— BOD: Obtained values for all samples ranged between 1.5 and 10.3 mg/l, and nine of them showed an oxygen concentration lower than 3 mg/l.

— Chlorite: Ten samples presented values higher than the guide level, varying between 26.3 mg/l and 49.9 mg/l.

— Ammonium: Six samples had results higher than the maximum permitted concentration, and they were between 0.6 and 16 mg/l.

(d) Point 4: This point is located at the gauging station E-64 in Santa Cristina d'Aro. The influence of a partially urbanised area. The stream channel at this point was totally dry during six of the sampling performances:

— Electric conductivity: Three samples presented values slightly higher than the guide level with values ranging between 407 μs/cm and 448 μs/cm.

— BOD: Obtained data varied between 3.3 and 8.6 mg/l. Seven of them reached after five days of storage an oxygen concentration lower than the limit.

— Chlorides: All samples showed values higher than the guide level. Results varied between 28.5 and 46.7 mg/l.

— Nitrate: This is the first downstream point which presents nitrate concentrations. However, all values are according to the suggested guide level.

(e) Point 5: It is located downstream the Water Depuration Plant in Castell d'Aro.

— Electric conductivity: All samples presented values higher than the permitted guide level, varying between 637 μs/cm and 1633 μs/cm.

— BOD: 100% of samples had an oxygen concentration lower than 3 mg/l, ranging from 0.3 mg/l to 0.6 mg/l. BOD was between 3.4 and 7.6 mg/l.

— Chlorite: All samples presented results higher than the suggested level, between 117.9 mg/l and 280.8 mg/l.

— Ammonium: All samples showed values higher than the maximum permitted concentration. They were between 4.4 mg/l and 109.6 mg/l.

— Sulphate: All samples presented results higher than the guide level (25 mg/l), varying between 86.4 and 215.6 mg/l.

— Nitrate: Obtained results are according to the suggested guide level.

(f) Point 6: This sampling point is located near the outlet of the Ridaura into the Mediterranean Sea in Platja d'Aro. Obtained results were similar than in point 5. From these results some remarks can be made:

Summarising, main point-source pollution in the Ridaura are point 2 (a small stream draining the Solius rubbish-dump), and point 5 (downstream of the water depuration plant). They both present highly disturbed chemical values for water quality. A more efficient treatment of the wasted waters by the Water Depuration Plant was observed during summer months. However, it is just in this season of the year when the Ridaura is dry and, thus, the aquifer is recharged with water from el Pasteral dam, used to complement the increment of the water consumption in the area. The pollution of the surface water of the Ridaura river could limit, therefore, the use of the river and the aquifer as a source for public supply. Moreover, its aesthetical look can affect the tourist development of the area.

Pollution of the Aquifer due to Wasted Waters

The main environmental problem related to wasted waters in the Ridaura drainage basin is the recharging of the aquifer during summer months with low quality water coming from the depuration plant, when the aquifer is not hydraulically connected with the water table.

The water depuration plant was built in 1983 for the municipalities of Santa Cristina d'Aro, Sant Feliu de Guixols, Platja d'Aro and Castell d'Aro, with an estimated population in the summer period of about 150,000 inhabitants. It is one of the largest plants in the Costa Brava with a depuration capacity of 3,900,000 m^3 of water per year. Monthly data of water depurated during the hydrological year 1985-86 shows a marked peak during summer months, especially in August, related to the increment of population (Serra, 1987).

An investigation carried out by the Escola Universitaria Politecnica de Girona (1984) showed an increment of phosphate, chlorite, ammonium downstream the water depuration plant. Similar results have been reported in this study. During summer months faecal streptococcus and coliforms were also detected. Since the river is dry in summer and early autumn, the discharge of the Ridaura during these months, downstream of the water depuration plant comes exclusively from the plant. Depurated waters are rich in nitrogen and phosphorous, which causes eutrofisation where the water remains calmed or retained.

Groundwater is used to irrigate the cultivated fields. Therefore a periodical control of this water should be done in order to verify its suitability for irrigation purposes. Some values are on the limits of the recommended values, for instance, conductivity, sulphate, ammonium, and chlorite. In addition, the greens of the Golf 'Mas Nou' are irrigated with wasted waters.

In order to avoid the recharging of the aquifer with polluted water and to prevent eutrofisation, two solutions were proposed (EUPG, 1984):

---- Construction of a submarine outlet from the plant to the sea.

—— Chanelisation of the lower reach of the downstream of the
plant, and refilling the streambed with gravel which filters the
water before it reaches the aquifer. First solution would be,
however, preferable, although it is more expensive.

Conclusions

The Ridaura drainage basin is defined by several typical
Mediterranean trends such as, seasonality of discharges, torrential
behaviour, high evapotranspiration rates (>0.85), and low runoff
coefficient (0.12) compared with similar areas in the Prelitoral
Mountains of the Catalan Coastal Ranges (Batalla and Sala 1993,
1996).

Dissolved load and water quality are characterised by the close
relation between the chemical composition of the streamflow and
the landuse. In particular, chlorite has a predominant position in
the water chemistry (25%) due to urban and tourist impacts in the
lowlands, but also due to atmospheric inputs and dry deposition in
the headwaters. There are several human, induced processes which
cause severe environmental degradation in the area, especially
pollution of the water due to the Solius rubbish-dump and
recharge of the aquifer with wasted waters from the water
depuration plant during summer months.

With regard to erosion, an increment of the specific sediment
yield with the catchment area has been observed. This pattern
differs from other of similar drainage basins in the Prelitoral
Mountains of the Catalan Coastal Ranges, where the sediment
yield decreases as the drainage basin increases, mainly due to
downstream sedimentation and the lower capacity of the
streamflow to transport. Thus, human activities, mainly
agriculture and urban works, appear to be capable of changing the
general pattern of sediment dynamics as observed in most drainage
basins.

Note

We are grateful for the support received by the EU from the Project EV5V-00043 *Desertification risk assessment and landuse planning in a Mediterranean coastal area*, and from the RESEL Project financed by the Spanish Ministry of Agriculture.

References

Batalla, R. and Sala, M. (1993): Balanc hidroquimic d'una conca granitica Mediterrania en estat seminatural. *Acta Geologica Hispanica*, 28,1.

Batalla, R. and Sala, M. (1996): Impact of landuse practices on the sediment yield of a partially disturbed Mediterranean catchment. *Zeitschrift fur Geomorphology*, Supp. 107, 79-93.

Environmental Monitoring and Support Laboratory: (1983): *Methods for chemical analysis of water and wastes*. Office of Research and Development, U.S. Environmental Protection Agency, Cincinnati, rep. 600-4-79-020.

ESCOLA UNIVERSITARIA POLITECNICA DE GIRONA (1984): *Estudi de la influencia de la planta depuradora del Consorci de la Costa Brava en la contaminacio del Ridaura*. UPB, Girona.

Martin, E.S. (1994): *Calidad del agua en la cuenca del Ridaura: variaciones espaciales y temporales*. Departament de Geografia Fisica i Analisi Geografica Regional, Universitat de Barcelona.

MINISTERIO DE OBRAS PUBLICAS, TRANSPORTES Y MEDIO AMBIENTE (1992): *Instrucciones de Carreteras*. Madrid.

Sala, M. (1979): *L'organitzacib de l'espai natural a les Gavarres*. Barcelona, Editorial Rafael Dalmau, 148.

Sala, M., Pernas, J., Ubeda, X., and Batalla, R.J. (1994): Studying downstream influence of landuse changes on runoff, solutes and sediment transport in a hilly Mediterranean coastal area. *Proceedings of the 3rd Conference on Assessment of Hydrological Temporal Variability Changes*. European Network of Experimental and Representatives Basins (ERB), Barcelona, 179-187.

Sala, M., Farguell, J. and Llorens, P. (2000): Runoff, Storm Runoff and
    Water Quality in Two Instrumented Small Catchments in the Catalan
    Coastal Ranges, ERB 2000 Proceedings.

Serra, M. (1987): Les depuradores de la Costa Brava, un cas insolit. *Revista
    de Girona*, 120, 42-48.

Trilla, M. (1980): *Estudi hidrogeologic de la conca del Ridaura.* Ajuntament
    de Santa Cristina d'Aro.

2

Current Environmental and Socio-Economic Degradation Processes in the Arid Andean Regions of North-Western Argentina

Dagmar Haase, Annegret Haase, Oliver Spott,
Lothar Linde and Thilo Weichel

The paper tries to assess, under a geographical and cultural point of view, the current state and forms of ecological degradation as well as the social decline in the arid Andean regions of north-western Argentina (province of Jujuy, Figure 1). It focuses on the historical development of both farming and pastoral agricultural systems in the Valley of Humahuaca and the Argentinian Puna as well as the current change in landuse caused by declining rainfalls during ENSO events, a highly variable rainfall intensity, problems of land ownership and the social and economical status of the indigenous Kolla population. Another issue of investigation is to show the problems and prospects of the future development of these rural, marginal and degraded mountain areas.

Introduction

Although north-western Argentina belongs to the most isolated areas of the country, the valley of Humahuaca and the northern

Figure 1
Situation of the Investigation Area of NW-Argentina

part of the Argentinian Puna can be characterised by a cultural
landscape with a long history full of changes (Haase and Haase,
2000). The region represents one of the most ancient settlement
areas of all Argentina. In pre-Columbian times, more than 5,000
years ago, the peoples of Atacama and Omaguaca developed
farming and pastoral agriculture in north-western Argentina. In
the 15th century these cultures were conquered and became part of
the mighty Inca Empire, (foundation of the El Pucara fortress near
Tilcara) consequently followed by a merger of the different ethnic
groups. The present province capitals of Salta and Jujuy were
founded after the Spanish conquest in the 16th century.

North-western Argentina is characterised by the following
landscape features: the Argentina, Puna in the north-west, so-called
"valles" (longitudinal valleys, e.g. Quebrada de Humahuaca, Valle
de Lerma, Valle Calchaqui), "nudos" (diagonal valleys) and little
"bolsones" (basins, cf. Figure 2, Roccatagliata, 1988, p.591). This
paper focuses on the Quebrada de Humahuaca and the northern
part of the Argentinian Puna (marked in Figure 2, cf also Figure 3
and 4).

Figure 2
Landscapes of Northwestern Argentina

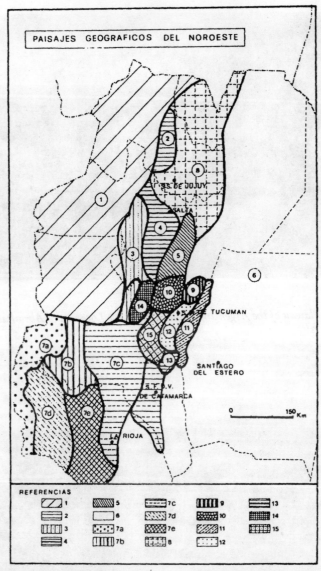

Source: *Roccatagliata, 1988*

(1 = Puna, 2 = Valley of Humahuaca, 3 = Valley of Calciqui, 4 = Valley of Lerma, 8 = Cordillera Oriental/Eastern Cordillera)

Figure 3
Tectonic Quebrada (Valley) of Humahuaca Near Tilcara (Purmamarca)

Figure 4
High Plateau of the Argentinian Puna (Near the Location of Abra Pampa)

The mountains of the Quebrada of Humahuaca (Valley of Humahuaca) are situated at an altitude of 2,000 to 3,000 meters above sea level and are based on volcanic ash layers and young sediments. In time, farming agriculture with its traditional cultivation methods (terraced farming, traditional irrigation systems) of the Kolla and Inca periods was replaced by other industries, e.g. mining. Today, former agricultural land lies abandoned thus reducing the cultivation area available in the mountains, which in turn increases the pressure on the landuse in the fertile valleys.

This degradation of the mountain ecosystems and the decline in the farming use of land are influenced by an enormous variability of climatic factors like rainfall intensity which differs considerably. As a consequence of these environmental conditions many local farmers were forced to reduce their herds of Lamas or Vicunas as well as cut down irrigation of their farming land. At the same time, a change in the landuse can be observed in north-western Argentina—the areas of farming land have been reduced to enlarge the areas for pastoral agriculture and breeding cattle, goat and sheep.

Together with the ecological degradation and economic changes lots of social problems are evolving. In addition to ecological facts the paper will explain the social backwardness of north-western Argentina and the exodus of especially young people taking place there. It also tries to find the underlying reasons in natural degradation processes, the change of the global climate (ENSO; Eckert, 1998) and socio-economic factors, e.g. the lack of alternative employment opportunities. As the region of the Humahuaca Valley suffers from growing marginalisation and pauperisation mainly, the younger population migrates to the cities or the Argentinian capital Buenos Aires.

The following paper mainly reflects on geographical and cultural approaches to show alternatives and prospects for the regional mountain economy and for the landuse to cope with the present problems. The article will give a detailed description and interpretation of the natural functions of the ecosystem, a theory concerning the possible regional influence of the global El-Niño phenomenon as well as a possible susceptibility of the surface material to soil erosion processes. Moreover, the reactivation of the traditional terraced farming as well as forms of regional tourism, such as hostelling, catering and the service sector as factors to cope with the poverty of the arid region will be mentioned.

Does the ENSO-Phenomenon Indicates Climate Variability?

The Valley of Humahuaca belongs to the diagonal arid, the dry transect, which covers large parts of the South American continent, mostly at the eastern side of the Andes. In the region of northwestern Argentina (Valley of Humahuaca and Puna ecosystem) steppe and dry or desert steppe ecosystems, tundra systems (local name: puna steppe) and full deserts have developed (Sal et al., 1993).

Since the beginning of this century precipitation has been monitored regularly. (Figure 5; data material: J.A.J. Hoffmann 1998): The station of La Quiaca is situated in the very north of the Province of Jujuy, 22°06' S, 65°36' W, 3442 m asl.

Figure 5

Variability of Precipitation (mm) in the Puna Region in the North of the Valley of Humahuaca

Station	1901-1910	1911-1920	1921-1930	1931-1940	1941-1950	1951-1960	1961-1970	1971-1980	1981-1990	Summer 1997
La Quiaca	334	299	324	268	381	319	305	330	371	98**
Average (literature*)	321	321	321	321	321	321	321	321	321	321
Difference (mm)	+ 13	-22	+ 3	-53	+ 60	-2	-16	+ 9	+ 50	(-223)

* cf Cabrera (1970): Territorios fitogeograficos de la Republica Argentina; **cf Information during a discussion at the INTA-station inin Abra Pampa

On the contrary to the Chaco Region or to the Pampa plains in the east and the southeast the precipitation of the Quebrada de Humahuaca and the Argentinian Puna often occur locally but not regionally. The distribution of the rainfall hardly depends on the

Figure 6

Temperature and Precipitation Dependency on the Altitude
(Source: Weichel, 2000)

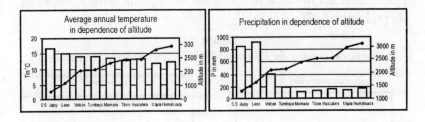

morphology of the sub-andean sierras (Hoffmann, 1971). Most parts of the annual precipitation occur in the summer period, the winter is characterised by aridity, salinisation and strong winds. Vargas-Gil and Culot (1998) mention for the arid Puna around La Quiaca a precipitation of 320 mm (Figure 5, 6) but an evapotranspiration of 567 mm. With temperature oscillations of 20-25°C (cf Nielsen, 1989) every day there occurs an additional strong insolation followed by an extreme surface evaporation of the superficial soil and sediment material. Contrary to the very arid Puna Region, the atmospheric humidity of the Valley of Humahauca and the other valleys (valles) is higher.

The variability of the climate parameters, most of all of the annual or seasonal precipitation combined with evapotranspiration are some of the limiting factors of agriculture and pastoral farming economy for the farmers of the Kolla people in the northwestern argentinian cultural landscape.

Through the screening effect of the surrounding mountains and the increasing altitude, the climate changes from humid tropical climate (Jujuy-Volcán) to the semi-arid -- arid desert climate (Humahuaca). For this reason the climate can be described as warm and dry. The annual precipitation between Tumbaya and Humahuaca is lesser than 200 mm, the surrounding mountains get more precipitation.

For the compilation of the influence of ENSO in northwestern Argentina the following material data on precipitation has been utilised, las Precipitaciones en el Noroeste Argentino (INTA, 1992), and data material of the ENSO-compilation by Schönwiese (1994; Figure 6, 7).

Figure 7

Boxplots Showing the Effects of ENSO in Northwestern Argentina

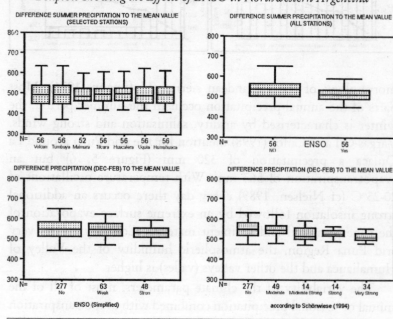

	ENSO intensity (simplified)	Events					
		valid		missing		total	
		N		N		N	
Difference annual sum	no	277	98,9%	3	1,1%	280	100,0%
(Dec-Feb) to the mean	weak	63	100,0%	0	,0%	63	100,0%
value (Dec-Feb) +500	strong	48	98,0%	1	2,0%	49	100,0%

Source: Spott et. al., 2000, Weichel, 2000

The summer months (main rainy season) are grouped so that December with the first two months of the following year formed an unit. After this the sum of precipitation was formed and evaluated with the program SPSS 7.5 for the 55 "summer". Next the mean value of all years was calculated. The difference between this mean value and the annual value create the measure for the difference.

For each station Boxplots were prepared to evaluate the difference of precipitation between the annual and the mean value, refer to the intensity of ENSO (Weichel, 2000).

The following conclusions could be made:

- All stations show a decrease of precipitation with increasing intensity of ENSO.
- Just the strong intensity of ENSO shows a considerable decrease of the summer precipitation, which is shown in Figure 7.

As the consequence of that there occurs the already described phenomena that agricultural land in the Quebrada and pastoral land in the Puna must be reduced. Moreover, the irrigation potential (reserves) is decreasing.

As further possible investigations towards forthcoming explanation and interpretation of the impact of global climate change on the regional climate of this arid region, the

- comparison with values of the, "La-Nina" cold water event as well as,
- a comparative investigation with other areas in Northwest Argentina to describe a possible regional differentiation between the Quebrada de Humahuaca – Punaplane – Precordillera could be aimed for.

Features of Landscape Development and Geomorphological Processes

The geomorphological situation of the Quebrada de Humahuaca and the Puna includes different kinds of erosion – accumulation and degradation-structures like the montanas (mountains), the bolsones (basins) and the valleys (valles) of the Puna region (Vargas-Gil and Culot, 1998). The geomorphology of this arid region can be characterised by steeply sloping up layers of mostly sedimentary paleozoic and mesozoic rocks. They are of volcanic origin in the western part and of marine origin adjacent eastwards. On this background several river terraces have been accumulated in different phases and have been eroded partially again. Beside current erosion structures like gullies, fluvial fans and debris surfaces spreading over the whole area, three W-E scree-fans of different size divide the investigated area in four parts. Beside these mesoscale structures, the area keeps a large amount of microscale forms like different kinds of salt crusts, secondary cemented fluvial sediments and sediment pyramids (Linde et al., 2000). Within the Puna basin meseta structures can be found.

The following descriptions about geological and geomorphological features within the valley are just related to an area between Maimara (2450 m asl) and Humahuaca (2950 m asl). They are based on fieldwork with special regard of geological, morphological and sedimentary conditions. One primary objective is a construction of simple models for a probable valley development during neogen and quaternary (Spott, 2000).

The main morphology of the valley belongs to a "basin and range" system and is deeply related to the tectonic structure as well as the geology of these Andean areas. So, regional tectonics and geology are the most crucial factors for the current occurrence of the Quebrada, whereas erosion seems to be just secondary and works only as a model for the interior valley surface. Important evidences for it are the SSW-NNE orientation of the valley, which is caused by the main orientation of regional fault structures (Cladouhos et al., 1994). Another hint is an obvious transition of a narrow and deep part of the Quebrada (1) around Maimara and Tilcara (outcrops are heavily faulted) to a vast basin area (2) surrounding Humahuaca (outcrops are scarcely faulted, building up a basin, Figure 8).

Within the main morphology caused by tectonic and geological structures we find a system of sedimentary remnants of

Figure 8

Geological/Morphological Sketches of the Quebrada-Valley near Tilcara (a) and Humahuaca (b, Spott, 2000).

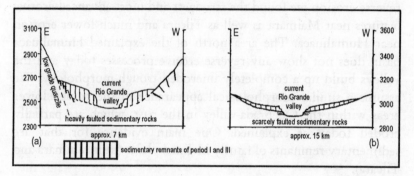

former as well as recent river and fan systems. So, the surface-structure of the interior valley today shows a rather complex construction of sediments (partial deeply eroded) built up in different periods. The appearance of the sedimentary remnants is obviously related to the named main morphology. The area surrounding Maimara and Tilcara (main morphology 1) shows mostly a heterogeneous system of quite different sedimentary remnants with huge thickness, steep slopes and currently deep erosion. The region near Humahuaca (main morphology 2), however, is mostly built up by basin-like structures of sedimentary remnants with faintly inclined slopes and fewer thickness as well as erosion (Figure 8).

Their formation is probably based on a continuous changing of periods dominated by erosion or by accumulation. Reasons for that are obvious phases of stronger or fainter tectonic activities (Cladouhos et. al., 1994) during the latest Andean orogenesis. So, accumulation of sediments dominates within the valley during phases of low tectonic activity. In times of increased tectonic activity the outlet-ratios of the concerned regions were changed through a partial lift up of different areas. This caused a relative lowering of the erosion base and led to a change from accumulating processes to mainly erosive processes which depends on an increased transport capacity of concerned outlet-systems after the lift up. The consequence is a reverse cutting into the prior accumulated sediments orientated from the erosion base to the remote areas of the catchment. According to these principle of reverse erosion we found the strongest and most advanced erosive features near Maimara as well as Tilcara and much lower erosion near Humahuaca. The area north of the explained Humahuaca basin does not show any reverse erosive processes today and the valleys build up a completely uneroded trough-morphology. We assume a similar morphological appearance of the named lower areas within the Quebrada-valley in the past, which are partially eroded today as explained. One main evidence for that are sedimentary remnants of faintly inclined plains (near Maimara and Tilcara).

Based on this scheme of further accumulating or further erosive phases and investigations of sedimentary remnants near Tilcara we classify four main periods of their formation as well as the development of the Quebrada-valley in this time. (i) The oldest and neogen (Amengual et al., 1974) sedimentary structures are simultaneous to the thickest accumulations. Their thickness range almost from the recent outlet-level of the Rio Grande (near Tilcara 2450 m asl) to heights of more than 2700 m asl) and above. A special characteristic for their occurrence is a clear and continuous change of layer's dominated by well-rounded gravel (few meters in thickness) and layer's dominated by silt as well as clay (1 to 2 meters in thickness). The roundness of the gravel indicates obviously an accumulation under fluvial conditions. Further more we recognised a faint inclination to the west against the western slope of the Quebrada-valley (tectonic effect after accumulation?; indicating a higher age?). Through the commence of these sediments near the recent outlet-level we know that it must have had existed prior to this (early) period. (ii) Followed by the period dominated by accumulation, we assume an erosive period caused by the explained tectonic effects where these sediments would be partially eroded. The evidence is the truncation of sediments of period I which are covered discordant by sediments of period III. (iii) During this period we had mainly the accumulation processes within the valley which built up the already explained plain remnants (inclined plains directed into the center of the valley). Their characteristics are of higher heterogeneous grain texture (no lamination) and a faint roundness of coarser grain sizes (local accumulation by fan-systems). (iv) The last period is characterised by mainly reverse erosive processes where sediments of period I and III would be eroded. So today we have a further ongoing process of reverse erosion within the Quebrada-valley. An interesting fact is the quite similar sediment structure of the current Rio Grade sediment and the sediments of period I which probably confirm to their fluvial origin.

Finally we have to consider the classified periods which are really simple imaginations or models for a probable development of the Quebrada-valley. Furthermore it has to be mentioned that other effects like climatic changes which are responsible for development of landscape could not be included in this study.

The Quebrada de Humahuaca is a tectonic valley with an enormous range of dry land geomorphology, concentrated on a comparatively small area. Main reason for the unique geomorphological features developed in this area are the geological backgrounds and the position of the Quebrada-valley within the range of special climatological phenomenon. The surrounding mountain range forms a natural barrier, hence this ecosystem is fenced off from most of the external influences. Due to the aridity characteristics of the valley, annual precipitation decreases from 300 mm (S) to 100 mm (N), day-night changes of temperature are large while the seasonal change of temperature is little. There is a heavy rainfall in summer. This together forms a valley that has an steep climatological gradient and therefore it is possible to study lots of ecosystems there, starting from moist forests up to desert level. This work specially focuses on the arid parts.

Important for the general geomorphology are the differences between the eroded material of various geological layers. The crystalline rock backwards delivers coarse gravel till the blocks which are with the red and yellow sandstone eroded to fine, sandy sediments. Both materials are mixed, different in the focused part and this displays the specific relationship between geology, relief and climatological features.

Based on this we have a geomorphological gradient from south to north that crosses four parts exemplary, the other four parts differ in a few parameters and can't be used for direct comparison. The southern part has, due to its lack of contact to the crystalline area, a dominance of sandy deposits and fine gravel. While the area close to the table rock is characterised by fans of coarse blocks, this material is going to be triturated on its way down. Large sediment pyramids can also be found. The adjacent sandstone area contains a wide range of erosion forms such as small soil pyramids covered by stones and different structures of (chemical) solution processes. With decreasing slope the eroded material was deposited as a large terrace. That accumulation process is finished. Currently this area is characterised by erosional forms which are initiated by runoff processes forming ephameral streams and along with it gullying, aeolic erosion and mass-wasting bring about further modification of the surface

The parts adjacent northwards display the gradient better due to their similar geomorphological features. This is a consequence of their large contact to the crystalline rock backwards. Hence they have a growing dominance of coarse materials which is in contrast to the previous described southern part. The area northwards can be split into three parts. The shape of these three parts is basin-like, open to the east. The size of the basins decreases to the north, which induces differences in the morphology of the structures. A feature every of these parts has, is a huge debris fan filling most of the basin.

The largest fan in the basin south, is structurally very similar to the previously described terrace, initially, there is a drained surface runoff, an exemplary debris surface and large gullies at the eastern end, all of which forms a fine sandy matrix including medium coarse gravel. Going northwards the next basin and its fan start to differ. Due to the change of geological and relief parameters there is an increasing transport capacity. This causes an increase of blocks and other coarse material, mostly of crystalline origin. Hence the fan is not very large, the matrix contains less fine material, the share of gravel increases and even blocks appear. There are deep fluvial drains on top of the fan, with no typical debris surface due to water runoff which is the dominating erosion form in contrast to the southern fans and terraces. Because of the character of the material, (inhomogenity, large amount of coarse sediments) gully-processes looses intensity. This gradient becomes clearest if compared finally with the most northern part. There the fan is extremely shallow and its edges go smooth to the surroundings. The fan-like structure or rather its origin is better seen in the satellite picture than in the field. Here the blocks and coarse erosives, mostly of crystalline origin, clearly dominate the geomorphological appearance, the share of fine grain sediments are reduced. Water runoff is most powerful here as compared with the basins described earlier. It influences particle size and composition. Due to that the morphology transforms itself from fan to braided river system with small, coarse terraces embracing several meandering riverbeds.

The investigation in the described area revealed a wide range of geomorphological forms of all sizes. Especially the large size forms show, there is the change in geomorphological shape, direction of slopes/gradient from ridge to valley. There is also a gradient along the valley representing other parameters like geological source and character of its erosives, material distribution and balance, transport medium and so on. Especially debris fans show along these gradient the change from fan-terrace characteristics to braided river dynamics in impressive examples.

Soil and Vegetation Factors

As different as the natural conditions and the water supply within the Valley of Humahuaca and the Puna are the different soils: good developed, deep soils within the valleys where the ground water level comes up near the earth surface (Fluvisols, Cambisols). These soils are intensively cultivated (potatoes, wheat, alfalfa; Vargas-Gil, 1998). Moreover, the valley consists of brown soils developed from gravel material. Today they are more or less drained. Towards the sierras soils, less developed A-and B-horizons occur (Regosols, Litosols, Aridisols). Around the "lagunas" (Lake of Pozuelos between Cochinoca and Rinconada) good developed and water drained soils can be found (Vargas-Gil, 1998).

On the surfaces of the mountain pediments there can be observed mostly quarternary aluvial-colluvial accumulations of fine material and gravel in combination with red arid soils (Aridisols; Linde et al., 2000).

The vegetation cover of the Quebrada de Humahuaca is formed by the cardones *(Trichocereus pasacana* and *terscheckii)*, little cactus forms like *Cassia crassiramca* and *Ceridium andicola*, floodplain gallery woods at the river sides (Q'eshwa zone; Nielsen, 1989) and jarilla-species (Nielsen, 1989), the Puna by *stipa-*, *artimisia-*, *poa-* or *festuca*-steppes (Sal et al., 1993). No tree vegetation within the Puna ecosystems does not exist.

Caused by the disappearance of terraced fields in favour of plain ones there can be observed different processes of top-soil erosion (aeolian and water forced erosion), as it was already described as morphological surface-process. Mostly at the edges of the slopes of the river terraces and the physically weathered and erodable sedimentary substratum, the valleys are in danger of being eroded very rapidly (observed in the Valley of Iruya). The missing vegetation cover (of the harvested fields and the arid environments) additionally supports the erosion processes caused by water and wind.

Such environmental degradation processes are increasingly linked with the existing poverty of population of the region, followed by phenomena like social and economic decline on one hand, structural weakness and employment alternatives on the other, which forces environmental degradation, land abandonment, landuse change and population exodus as well. This "circulus viciousus" will be explained in the next paragraphs (cf., Figure 9).

... riqueza humana ya que no económica: the socio-economic sphere

Northwestern Argentina represents one of the weakest regions of the whole country due to its economic development and the social impoverishment of the population. From the INDEC (Instituto Nacional de Estadistica y Cencos) it was declared a "zone of regional poverty/backwardness" (compare Sarrailh et al. 1998, p.110). The marginality of the region becomes obvious by its rare settlement, only 1% of the population of the whole country lives here. Today, within the northwestern argentinian region there are about 1.13 inhabitants per sq km (Sal et al., 1993).

In the Argentinian Puna and the Valley of Humahuaca tiny villages or some small towns like Humahuaca itself (4,000 inhabitants), Tilcara (2,200), in the Quebrada as well as La Quiaca (9,000, border town to Bolivia), Cochinoca (7,500) and Abra Pampa (3,000) in the Puna region form the settlement network.

Over centuries the economic development of Northwestern Argentina was orientated to meet the needs of the colonial authorities. It was aimed at the providing the Spanish mother country with resources and goods and it contributed to the massive exploitation of the natural and human potential of the region (Roccatagliata et al., 1988, pp. 594-595). As a consequence, the specialised terraced agriculture (Merlino, 1976-80) as well as the breeding of lamas which represented these territories up to the Spanish conquest declined. Instead of this, the mining, first of all of gold and silver, and the cultivation of tobacco and sugar cane at plantations (Rivero, 1990) had to be developed.

After the independence of Argentina in the 19th century, the northwestern region could not wipe aside its backwardness/ underdevelopment. Neither was there a fundamental reconstruction of the economic base which meant agriculture, mining, local trade and handicraft from the colonial period took place, nor were there settled bigger investments in the building up of an infrastructure for regional industries or the development of the road and railways. The economic problems are reflected even today in social features as a high rate of underemployment, a bad education level of the majority of the population as well as a high rate of school dropouts up to 50%, and a low life expectancy of the people, in the rural areas and in the slums of the bigger towns. The core problem is the lack of employment, especially outside the agriculture and mining sectors. Thus, the emigration process of young people from the andean villages and from the Puna region has been continuing till today (cf., Figure 9).

Economic Problems: Employment Decline in Agriculture and Minery—The Lack of Alternatives

The agriculture of Northwestern Argentina consists of farming, gardening horticulture and stockbreeding extensive pastoral economy. The areas under cultivation (farms) are situated mainly in the valleys, e.g. the Quebrada of Humahuaca, up to an altitude of 3,000 meters, asl. The most important plants being cultivated are legumes (e.g. beans, quinoa, kiwicha), potatoes and various kinds of chili pepper, additionally maize, fruits, vegetables or export plants like tobacco, coffee and vine in lower altitudes to 2,500 meters (Sarrailh et al., 1998, p.101). Generally speaking, only a very small percentage of the surface of the arable land (3 to 10%) of the whole Northwestern Argentina can be used for agriculture (Roccatagliata, p. 593). The level of mechanisation of the farms is very low and the main gadget even now continues to be the hoe. The pasture lands register the lower percentage of cattle than that of goat and sheep.

The current land-use in the Quebrada of Humahuaca consists of irrigated agriculture, the irrigated land covers an area of about 380.000 ha in Northwestern Argentina. Contrary to the past landuse, farmlands decreased in favour of pastoral lands, not only in the Puna Region but also in the fertile Valley of Humahuaca. During an expert interview with scientists from the INTA station in Abra Pampa (Puna Region) it was found that there could exist a certain influence of the ENSO-phenomenon on the precipitation variability and remarkable loss of rainfall in the Puna and Quebrada Region during the summer period 1997.

The alluvial terraces, floodplains and creek valleys of the main Rio grande – River of the Quebrada de Humahuaca are intensively used for farming. Furthermore, the farmers utilise the water of the arroyos (erosion streams, creeks, gullies) coming from the Andes in the west to irrigate their land. Today commercial agriculture only exists in the Valles de Calchaquis, but in the Valley of Humahuaca and the Argentinian Puna subsistence agriculture dominates. The Puna Region in the north of the Province of Jujuy is utilised mainly by pastoral agriculturists (lamas, vicunas, sheep, goats). The pastoral farming land covers an area of about 8,550.000 ha.

The problem of current land-use-change is caused by the (e.g. described by Otonello, 1982; Haase, D., 1999) decreasing areas of agricultural land and increasing areas of pastoral land. Forced degradation of soil and fine sediments indicated by rising soil density and superficial water and wind erosion (processes described by Opp, et al., 1999). Moreover, the former traditional farming systems (terraced farming of the pre-Inca and Inca periods) were able to protect the soil from the main forms of erosion in form of stone walls around the fields and accumulation of humic and plant material at the surface of the terraced fields during the restoration periods. This is mostly missing today.

The described landuse change means that less labour opportunities exists constantly in the agricultural sector in Northwestern Argentina. The preference of (extensive) stockbreeding instead of farming leads to a decline of the small scale farming at the slopes of the valley of Humahuaca. As a consequence, during the last decade less people found work in this field. Furthermore, the regional agricultural production of Northwestern Argentina could not compete with imported good products.

Respecting the labour market, an alternative to the agricultural sector consists in the mining industry which represents an important basis for the regional economy and labour market in the NOA. There are produced in the mines of El Aguilar first of all silver, lead, tin and zinc, partly in opencast mining. Generally speaking, here are exploited 90% of the metallic resources of Argentina. A part of the production goes as export goods to abroad. Up to the last years, in the mines worked several thousands of miners. Since 1997/98, however, the mining in Northwestern Argentina led to a decline; therefore lots of workers had to be dismissed from their jobs (information from an interview with a member of the NGO API in Tilcara in March, 1998).[1]

The Social Consequences: Marginalisation and Exodus of Population

Roughly speaking, the decline of labour opportunities in the agricultural sector leads to a rising exodus of people in the age groups capable to work and with a higher education from the mountain regions to the province capitals like Jujuy and Salta, or directly to the country's capital, Buenos Aires. During the last years, as it was mentioned above, the mining towns like Aguilar are suffering from a economic repression, and it can be supposed that this will initiate a emigration wave to the urban centres, too *(Han, 2000, pp 128-141).*

The said cities are characterised by a remarkable population growth as a consequence of the migrations. So, for instance, the city of Jujuy shows the biggest rate of population growth of the whole Argentina (69.8% per year). The migrants mostly get bad paid jobs in the informal sector. The majority of them lives in the slums or in squatter settlements at the edges of the towns (in Argentina they are called villas miserias). The miserable hygienic conditions of these dwelling places cause the bad health estate of the major part of the migrants.

In the Andean mountain areas of Northwestern Argentina, the lack of labour opportunities and the exodus of active generations lead to the impoverishment of the settlements and their population. There are left so-called "villages of children and elderly people" like Tres Cruces, not far from Abra Pampa. In these villages, the deformation of the demographic and social structure of the population leads to a decline of the family life. The lack of any social assistance (social network with assistance for unemployed people, pensioners, families with 3 and more children, single mothers with children, handicapped persons etc.) is responsible for high death rates among children, the occurrence of epidemics (e.g. cholera) or a wide-spread malnutrition.

Taking into consideration the described problems of environmental degradation and socio-economic decline, there is a question whether the region of Northwestern Argentina also in future will suffer from backwardness and poverty, though it represents one of the oldest cultural landscapes of South America with still living indigenous traditions.

Hence, the following facts reflect the potential of the region to be utilised for its development:

Figure 9

Interrelations between Geofactors and Socio-economic Development in Northwestern Argentina

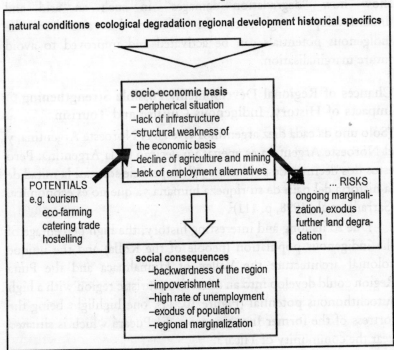

natural conditions ecological degradation regional development historical specifics

socio-economic basis
– peripherical situation
–lack of infrastructure
–structural weakness of the economic basis
–decline of agriculture and mining
–lack of employment alternatives

POTENTIALS
e.g. tourism
     eco-farming
     catering trade
     hostelling

... RISKS
ongoing marginalization, exodus further land degradation

social consequences
–backwardness of the region
–impoverishment
–high rate of unemployment
–exodus of population
–regional marginalization

Layout: A.Haase/D. Haase 2000

- Up to day, a lot of natural and socio-economic potentials of the region are not yet used.
- The future of the region will at least depend on the adequate combination of the development of the physical geographical basis and the socio-economic requirements of the inhabitants.
- Thereby the indigenous potential of the region (natural resources, cultural traditions, regional knowledge, borderland situation) may play an integrative role.

One development perspective for north-western Argentina seems to be tourism. Therefore, the question will be looked upon how, tourism can contribute to long-term regional development in order to strengthen the socio-economic basis of the people as well as to protect the natural environment to prevent degradation risks.

For that reason the following part of the paper will not only show how "degradation damages" in such an arid and underdeveloped region can be reversed, but also how the indigenous potentials can be activated and improved to avoid future marginalisation.

Chances of Regional Development: Possible Strengthening Impacts of History, Indigenous Culture and Tourism

"Sólo uno da cada diez argentinos vive en el Noroeste Argentina, y el Noroeste Argentina es apenas un decimo de la Argentina. Pero estos dos decimos-de poblacion y de superficie-son casi la mitad de la historia del pais: de su riqueza humana ya que no es economica. (Sarrailh et al., 1998, p. 111)[2]

Due to its long and interesting history, the high percentage of the indigenous population (people of the Kolla) and the unique colonial architecture the Valley of Humahuaca and the Puna Region could develop into an attractive touristic region with a high autochthonous potential (Figure 10, 11), one highlight being the fortress of the former Inca-emperors, El Pucará which is situated near the community of Tilcará.

The local and regional potential of the Valley of Humahuaca and the Puna Region is really high for its wonderful colourful landscape with red, brown, yellow and green rock formations, the indigenous culture of the Kolla people, which is unique in Argentina, old churches with preserved paintings of the well-known School of Lima. These facts are mostly attractive to foreign tourists who come from the province capitals of Salta and Jujuy for a longer stay or a short trip. Up to now only a few opportunities exist for tourists. For the future, attention should be paid to the fact that not only new hotels and restaurants as the commercial and regional basis should be planned but also lodging possibilities could be offered by rural families in their homes.

Vorano *(1997)* describes one possible form of tourism, the so-called agro-tourism. This kind of sustainable and nature-connected tourism was also mentioned in the expert interviews by the local responsibles of different non-governmental organisations[3] (Radio Pirca, API, church initiatives) as a chance to develop and to protect this remote region and its fragile ecosystem. Moreover, this could also be a form of sustainable economy for a formerly agricultural region and an alternative approach to develop an isolated and arid region. Furthermore, agro-tourism has been more and more developing all over Argentina during the last decades.

This kind of agro-or ecological tourism has many advantages for both the tourists on one hand and the local population on the other: a local service producing a local source of income for the families. Moreover, a loss of taxes and money paid to the province tourism agencies and province governments can be avoided. Furthermore, the families are given the opportunity to sell locally produced handicraft, natural products, bread, jam, potatoes, woolen shirts and sombreros to the tourists. This system of developed agrotourism supports a better income of the local population which finally leads to a better potential for the farming agriculture in general. Moreover, the necessity to cater for the tourists staying in the families, to invest in the farm land and to preserve the soils as a vital production base arises. As a matter of

Figure 10
The Inca fortress "El Pucará" in Tilcará

fact the areas with autochthonous, traditional crops like *Quinoa* or *Kiwicha* which produce indigenous food for the tourists and grow easily in this fragile ecosystem with a highly variable climate can be enlarged. This may be seen as a chance to avoid the total change in landuse from farming to pastoral agriculture (Figure 9).

As for the population of the Kolla people, who suffer from a high degree land exodus of the mostly young they could be given the opportunity to work personally in the field of tourism, i. e. as a cook, lodge owner, local guide for history and traditions or local (regional) travel adviser. Educational programmes for the local population would be necessary as well as investments in the infrastructure of the Valley of Humahuaca and the Puna Region (water supply, canalisation, traffic routes).

Today tourism in north-western Argentina is growing, however slowly. Most of the foreign tourists visiting Argentina prefer the mountains and glaciers in the south of the country and the capital of Buenos Aires and avoid the dry and poor north.

Figure 11
Push and Pull Effects in Northwestern Argentina

Push and pull effects of a peripherical cultural landscape: potentials and problems of the region of the valley of Humahuaca and the Argentinian Puna

	INDICATOR	CONSEQUENCE
Pull	landscape	interest, photo-tourism
	indigenous history, historical past	
	ancient churches	historical interest
	archaeological excavations	
	ecological farming	"Eco-Freaks"
	adventure	adventure tourism
	mountains	climber, sportsman
	image	cultural tourism
Push	lack of working places	migration, land-exodus
	poverty	migration, less of funds for initiatives
	climate variability	limiting agriculture farming
	natural conditions, pastorism	land degradation
	missing ownership of land	delay of local/regional investments
	periphery/marginality	poverty, passivity
		low level of education
	image	"poorhouse"

Layout: D. Haase/A. Haase 2000

Nevertheless, some examples of new and ecologically built youth hostels in Tilcara and Humahuaca exist (wood of the cardones was used as building material). Historicl artefacts of the pre-Inca and Inca-period of north-western Argentina are on display in local museums in Pucara, Tilcara and Humahuaca.

Nowadays local organisations like the Radio channel FM PIRCA, the movement "Viltipoco" or the NGO-project API (Asociacion para la promocion integral in Tilcara) deal with the regional development in the form of development initiatives which include participative approaches, support of local women, the farmers and their families by means of common agricultural work as well as the foundation of kindergardens including primary education for children and the support of the role of women within the rural society and family. Moreover, educational programmes in agriculture, traditional cultivation methods (terraced farming), garden subsistence cultivation, managing the commercial sale of the agricultural products and, finally, educational programmes on environmental protection were initiated. Since 1995 also governmental initiatives deal with the co-operation of scientists and farmers, credits and animal care, e.g. the Programma Social Agropecuario 1996, National Institute for Technology and Agriculture INTA, Buenos Aires.

Conclusions

This paper discussed geographical or geo-ecological and cultural aspects to explain and to interpret the processes of environmental degradation, the effects of climate change, the current change in landuse and the linked socio-economic problems in north-western Argentina, mainly in the Valley of Humahuaca and the Puna Region. It was shown that only on the basis of exact scientific data material in combination with "soft" factors like historical and regional traditions and customs it will be possible to protect the nature and the traditional economy as well as to combine it with new, perspective branches like tourism (e.g. agrotourism).

Tourism often seems to be a soft impact but it can be developed in a sustainable way. The most important fact is to activate the local potential of nature and the people, to enlarge their capabilities and to save the environment together with their economy.

To summarise it can be mentioned that

— regional development concepts for the arid and marginal region of north-western Argentina have to include aspects of sustainability and flexibility,

— development concepts for landscape management have to include traditions and the "human capital" of the population,

— tourism has to be understood as a new economic branch and source of income for the population,

—— tourism is able to strengthen the local and regional infrastructure as well as the administrative and decision-making structures which also support the regional actors.

Last but not least, all future activities to improve the economical basis should respect the specific regional environmental, socio-cultural and cultural characteristics of north-western Argentina.

Notes

1. The Northwestern region moreover represents an exploitation area for oil and natural gas. The main production places, however, are situated at the eastern slopes of the Andean mountains or in the subandean valley near the town of Tartagal. In 1994, there were produced 10% of the argentinian natural gas output (3.141.000 m^3) and 5% of the oil (966.000 m^3), compare *Sarrailh* et al. (1998, pp.103-104).

2. Only one-tenth of the population of all Argentina lives in the north-western part of the country. And north-western Argentina represents only 10% of the country's territory. However, these 10% represent half of the country's history and wealth, even if no longer in the field of economy. (translated be D. and A. Haase).

3. Interview with Radio PIRCA in April 1993, Tilcara, Interview with API in March 1998, Tilcara, Interview with catholic initiative OCLADE in Humahuaca in March 1998.

References

Amengual, R. and J. C. M. Zanettini (1974). Geologia de la Quebrada de Humahuaca entre Uquia y Purmamarca (Provincia de Jujuy). In: Revista de la Asociación Geologica Argentina. Tomo XXIX/No 1. Jujuy.

Bianchi, A.R. (1992): Regiones productivas de Salta y Jujuy, Panorama Agropequario, INTA, pp 9-14.

Bianchi, A.R., C.E. Yanez (1992); Las Precipitaciones en el Noroeste Argentino (Segundo Edicion), INTA Secretaria de Agricultura, Ganaderra y Pesca de la Nacion. EEAS-Centro Regional Salta Jujuy.

Censo Nacional de poblacion y vivienda (1991): Instituto Nacional de Estadistica y Censos, Buenos Aires 1991.

Cladouhos, T.T; R.W. Allmendinger; B. Coira and E. Farrar (1994). Late Cenozoic deformation on the Central Andes: fault kinematics from the northern Puna, northwestern Argentina and southwestern Bolivia. In: Journal of South American Earth Sciences. Vol. 7/No. 2.

Eckert, CH. (1998): El Niño, Munchen.

Haase, D. and A. Haase (2000): Geographical and cultural approaches towards the assessment of traditional mountain land-uses in Northwestern Argentina. *Profil* 18:26, Stuttgart

Haase, D. and A. Haase (1998): Ecosystems of the andean mountains and the Argentine Puna of Northwestern Argentina-landscape, it's exploitation and conflicts. In: Terra *Nostra* 5:61-62.

Han, P. (2000): Soziologie der Migration, Stuttgart.

Hoffmann, J.A.J., S.E. Nunez, W.M. Vargas (1997): Temperature, humidity and precipitation variations in Argentina and the adjacent sub-antartic region during the present century, Meteorolog. Zeitschrift Stuttgart, N,F. 6, 3-11.

Hoffmann, J. et al. (1994); Die Niederschlagscharackteristik in Nordwest-und Zentralargentinien und ihre wirtschaftliche Bedeutung, *Geökodynamik* Bensheim.

Hoffmann, J.A.J. (1971): La distribución geografica de las precipitaciones en el noroeste argentino. Meteorologica, Vol.ll, No.1,2,3, Buenos Aires.

INTA (1996): Programa Social Agropecuario, Buenos Aires.

Linde, L., Spott, O., Weichel, T. and D. Haase (2000): Geomorphological processes in the Quebrada de Humahuaca (NW-Argentina), Abstract for the Latin Amerika Symposium, Stuttgart.

Meilán, D.; H. Nielsen; R. Page and J. Mendia (1998). Estudio Geologico integrado de la Quebrada de Humahuaca Geologia Regional y Geomorfologia. Instituto de Geologia y recursos Minerales. Jujuy.

Merlino, R.: Pastoreo y agricultura en el altiplano meridional. RUNA, Part 1-2, Buenos Aires 1976-80,13:113-119.

Nielsen, A. (1989): La ocupación indigena del territorio Humahuaca oriental durante lod periodos de desarollo regionales e Inka, Córdoba.

Opp, CH., Haase, D. and V. Khakimov (1999): Soils and soil degradation in the Tuvinian part of the Uvs-Noor Basin. Proceedings International Conference, Role of soil in functioning of ecosystems" Sept 7-10, 1999, Lublin, pp 467-468.

Otonelo, M. (1982): Environment, human settlement and agriculture in the Puna de Jujuy (Argentina): A case study of land-use-change. Mountain Research and Development: Vol. 2, No.1, United Nations.

Rivero, J.O. (1990): Puna, zafra y socavon. Madrid.

Roccataglitata, J.A. (1992): La Argentina: Geografía general y los marcos regionales, Planeta Buenos Aires.

Sal, J.D., Vargas-Gil, J.R., Nieva, I.J., Tubello, D.A.G. and A.J. Alvarez (1993): Producción y utilización de los pastizales andinos de la república argentina. INTA, proyecto apoyo del centro internacional de investigación para el desarollo del canada (CIID).

Sarasola, C. (1986): Nuestros paisanos los indios. Buenos Aires.

Sarrailh, E.E.O. et al. (1998): la Argentina y sus espacios geográficos, Buenos Aires.

Schonwiese, C.-D.(1994); Klimatologie, UTB Ulmer Stuttgart.

Spott, O., Linde, L., Weichel, T. and D. Haase (2000): Geological, geomorphological and climatological features of the Quebrada de Humahuaca. Proceedings of the 13th Int. Kolloquium of Latin America, Stuttgart *(in press)*.

Vargas-Gil, J.R. (1998): Mapa de capacidad de uso de las tierras del extremo norte de la puna argentina, INTA.

Vargas-Gil, J.R. and J.P. Culot (1998): Mapa de reconocimiento de los suelos del extremo norte de la puna argentina, INTA.

Vorano, A.E. (1997): Agroturismo: Una posibilidad para armonizar el desarollo con la convervación ?, Panorama Agropequario, 1997, pp 17-18, INTA Salta.

Weichel, T. (2000): Der Zusammenhang der Niederschlagsmenge in Abhängigkeit von ENSO-Ereignissen an ausgewählten Stationen in der Quebrada de Humahuaca im Nordwesten Argentniens, Praktikumsbericht.

Zuleta, S.V. (1984): El rostro de Humahuaca. Humahuaca.

Zuleta, S.V. (1988): Humahuacamanta. Humahuaca, Jujuy.

3

Land Degradation:
A Menace that Haunts Humanity

R.Y. Singh

The United Nations (UNEP Report, 1992-III / 3 October, 1991) considers land degradation and desertification as synonymous terms.

The Rio de Janeiro Earth Summit (1992) formally defined desertification as 'Land degradation in arid and semi-arid and dry sub-humid areas resulting from various factors including climatic variations and human activities' (UNCED, 1992).

It is also true that:

> Failure to use consistent assessment criteria and consistent definitions of what comprises a dryland has caused regrettable confusion (Williams and Balling, 1994). The concept of desertification, however, has now been enlarged to allow a place for the adverse impact of climate desiccation and a prolonged or more frequent droughts (Williams and Balling, 1994).

The agreement reached at the Earth Summit in Rio de Janeiro helped in finalising the Convention on Desertification signed by 87 at first, and later by 115 countries in Paris on 14-15 October, 1994, to combat desertification and more than 75 of these have ratified it.

The aim of the paper is to look into causes and consequences of desertification and to highlight a few measures to mitigate it in the light of studies and reports made earlier.

The Causes and Consequences of Desertification

It is quite reasonable to think that the basic cause for the development of the desertification process is human impact. The problem of desertification hinges upon a wide spectrum ranging from welfare to survival of human species. The causes and resulting consequences would be better summarised (see Grainger, 1990; Bojo, 1991; Drenge, Kassas and Rozanov, 1990; Scoging, 1993) in relation to landuse, government policies adopted, as well as natural phenomena and their interactions. The factors related to landuses, i.e., overcultivation, overgrazing, mismanagement of irrigated lands, deforestation and fire based shifting cultivation affect physical processes, e.g., depletion of soil nutrients and organic matter, decline in soil structure, water permeability, susceptibility to erosion, compaction of soil, sand dune mobilisation, loss of bio-diversity, increased run off, waterlogging and salinisation of soil, loss of soil stabilising vegetation, increased frequency of dust storms and promotion and growth of unpalatable woody shrubs in place of medicinal herbage in burnt areas.

Secondly, the government policies pertaining to population planning, irrigation subsidies, land tenure and settlement, infrastructure like roads, dams, canals, boreholes, etc., promotion of cash crops, price increase in some crops, conflicts and refugee-settlements and high interest rates create over-exploitative landuse practices, exacerbate flooding and salinisation, promote concentrated use of land by nomads/refugees exceeding the carrying capacity, enhance lowering of water table, problem of silting and displacement of people and subsistence cropping, marginal landuse, less resilient mono-culture, migration, forced grazing and cultivation beyond land capacity level. Thirdly, occasional droughts decrease vegetation cover, increase land vulnerability for soil erosion and create an environment which exacerbates over exploitation.

Darkoh (1989, pp. 62-71) attributes a number of impediments to past antidesertification measures related to planning/policy, resource use, people's participation and monitoring as enumerated below:

(i) lack of political will regarding National plans addressing control measures in development planning, (ii) lack of prioritisation, (iii) inadequate identification of social and economic causes, (iv) poorly designated conservation works and over-emphasis on technical solutions, (v) lack of voluntary/allocated funding for United Nations in fulfilling the Plan of Action, (vi) inadequate human resources, i.e., semi-skilled or skilled personnel, (vii) inadequate scientific machinery for land management and assessment, of desertification, (viii) little account of the needs, skills, experience, wisdom and aspirations of the affected people, (ix) imposing management upon people who may not willingly accept external interference, (x) lack of coordination and integrated approach in government departments, (xi) lack of regular monitoring mechanism and involvement of people to combat desertification, (xii) lack of close supervision and control of grazing to ensure a proper balance between numbers of livestock and useful plant cover, (xiii) introduction of technology without prior consideration of all factors in the environment, (xiv) lack of follow-up observations on management impacts, and (xv) failure to combine efforts of United Nations agencies for multi-disciplinary approach at all levels.

Otinda has rightly observed that:

Compounding the land degrading effects have been the inequitable international trade, international debt and Structural Adjustment Programmes (SAPs) (Otinda, 1997, p. 3).

The overview prepared by the Conference on Desertification emphasises that its impact is generally attributable to objective factors like population growth and corresponding increases in needs leading to intensification in the use of natural resources. Also it includes environmental hazards such as drought. The symptoms of the process of desertification or its end-result include (i) reduction of yield or crop failure in irrigated or rainfed farmland, (ii) reduction of perennial plant cover and bio-mass produced by rangeland and the consequent depletion of food available to livestock, (iii) reduction of available woody biomass and the consequent extension of distance to resources of fuel wood or building material, (iv) reduction of available water due to decreasing of river flow or groundwater resources, (v) encroachment of sand that may overwhelm productive land, settlements or transport and communications systems, (vi) increased flooding, sedimentation of water-bodies, water and air pollution, (vii) social disruption due to deterioration of life-support systems that calls for outside help (relief aid) or that prompts people to seek haven elsewhere (the phenomenon of environmental refugees) (UNEP Report, 1992, p.3). The problem becomes acute in relation to following consequences:

Socio-economically it is the main cause of global loss of productive land resources; it causes economic instability and political unrest in the effected areas; it extends enormous pressure on the economy and stability of societies outside; it prevents the achievement of sustainable development and; it directly threatens the health and nutrition status.

Environmentally it is an increasing element of planetary environmental degradation; it contributes heavily to loss of global bio-diversity; it increases the loss of biomass and bio-productivity particularly reduction in carbon dioxide sink; it contributes to global climate change by increasing land surface albedo. As a global problem desertification needs a coordinated effort at the international level.

An overriding socio-economic issue in desertification is the imbalance of power and access to strategic resources among different groups within society. Thus, desertification has been one of the main factors in the migration of subsistence farmers and pastoralists. In most of the developing countries/world, population pressure and inappropriate farming practices contribute to soil impoverishment and erosion, rampant deforestation, overgrazing of common lands and mis-use of agro-chemicals. Other factors include unequal distribution of land, the lack of agricultural extension services to subsistence and small scale farmers, limited access to credit and better technologies and faulty agricultural policies that favour export crops over food crops.

Assessment of Desertification

The principal conclusion of the overview of UNCOD, 1977, is that desertification processes (which reduce the biological productivity of arid lands) are currently developing everywhere throughout the world at an accelerating rate. The rates at which lands are withdrawn from agriculture as a result of desertification is nearly fifty thousand square kilometers per year. "If we consider that according to the World Map of Desertification the total area of potentially productive arid lands threatened by desertification is 45 million square kilometers, that is 30% of the entire earth's surface and that it currently sustains 600-700 million persons, the scale of the problem and its socio-economic significance becomes apparent" (Gerasimov, p.129).

According to soil/vegetation data world drylands constitute 6.45 billion hectares or 43% of global land. According to climatic data, world drylands constitute 5.55 billion hectares or 37% of global land. The difference of 0.9 billion hectares or 6% might be attributed to man-made deserts. The area threatened at least moderately by desertification within the drylands amounts to 3.97 billion ha or 75.1% of the total drylands, excluding hyper-arid deserts. More than 100 countries are affected which inhabit more than 15% of the world population. The population of areas suffering by severe desertification total 78.5 million people. The annual rate of degradation in arid and semi-arid areas alone amounts to a total of 5.25 million ha of which 0.125 million ha, 2.5 million ha and 3.2 million ha were covered by irrigated lands, rain-fed croplands and rangelands respectively.

Another assessment of the status of desertification was undertaken by UNEP in 1981. It showed that desertification has continued to spread and intensify, and land degraded to desert-like conditions continued to increase at the rate of 6 million ha annually, while land reduced to zero or negative net economic productivity was showing an increase of 20 to 21 million ha annually. Areas affected by at least moderate desertification were 3100 million ha of rangelands, 335 million ha of rainfed croplands and 40 million ha of irrigated land, totaling 3,475 million ha.

The most recent assessment was made in 1991 and a new map of drylands was prepared by Global Environmental Monitoring System/Global Resource Information Data Base Programme Activity Center of UNEP (Figure 1).

Figure 2 shows the areas with risks of desertification which threatens nearly one third of the Earth's land surface, affecting the lives of, at least, 850 million people. During the period 1978 to 1991 desertification had lead to migration of subsistence farmers and pastoralists to the slums and shanty towns of major cities. The report states that in 1984 and 1985 an estimated 30 to 35 million people in 21 African countries were severely affected by droughts. Approximately 10 million were displaced and became known as environmental refugees. While the burden on farmers and pastoralists in these countries can be traced partly to international policies and markets, it also has roots in local land tenure practices and usufruct rights as well as domestic priorities, that often favour the urban consumer over the rural producer. The marginalised people received little support in breaking the vicious cycles that mismanaged land. High growth rates and high rates of urbanisation have sucked the steady stream of soil nutrients (in the form of food, fuel wood and charcoal). This rapid transition from rural to urban societies has not been matched by an equally rapid replenishment of soil nutrients. Demands of production have pushed the limits of production into increasingly marginal lands which are much more susceptible to processes such as erosion and salinisation. Agricultural expansion to marginal lands frequently resulted in rapid land degradation.

Beside affecting the socio-economic aspects the menace has clear impact on the state of the land. The two data sets produced by ICASALS and ISRIC and UNEP in 1990 speak about land and

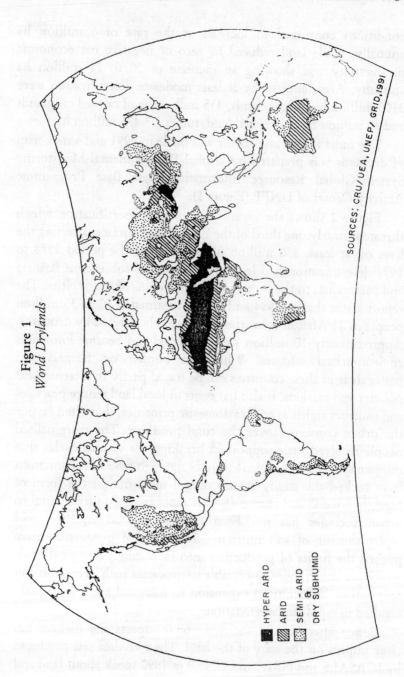

Figure 1
World Drylands

SOURCES: CRU/UEA, UNEP/ GRID, 1991

HYPER ARID

ARID

SEMI - ARID
&
DRY SUBHUMID

Figure 2
Desertification

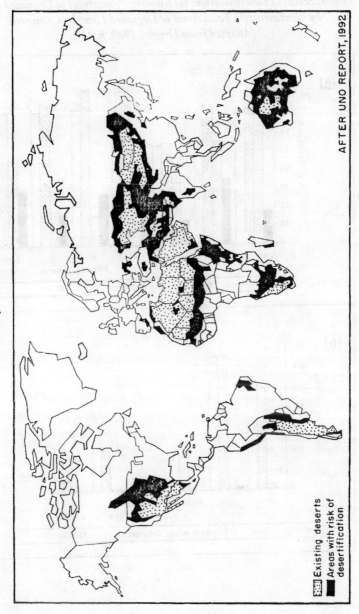

AFTER UNO REPORT, 1992

⠿ Existing deserts

■ Areas with risk of desertification

Figure 3
Global Status of Desertification: (a) Relative Proportions of Degraded Land
by Continent; (b) Total Areas of Degraded Land by Continent,
(Adapted from Dregne, 1989, p. 13).

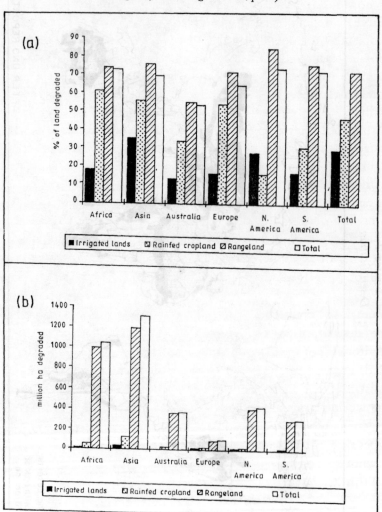

soil degradation respectively. The total area of drylands affected by degradation at present comes to nearly 3.6 billion ha or about 70% of the total dryland.. Soil scientists have established that the world is now losing about 1.5 million ha of irrigated lands annually (UNEP, Report, 1992, p.16). Nearly 216 million ha of rainfed croplands or about 47% of their total area in the world's drylands (457 m. ha) are affected by various processes of degradation. Rainfed cropland in the drylands constitutes nearly 36% of the total area in the world (1260 m.ha). It was estimated that 3.5 to 4.0 million ha of rainfed croplands are currently lost every year throughout the world's dryland. The percentage of degraded rangelands has similarly shown largest area in Asia and Africa. The figure presents an increase of some 233 million ha in comparison with the 1984 assessment, approximately 7.5%. The annual losses of rangelands within the drylands amounts to 4.5 to 5.8 million ha. Thus 70% of agriculturally used drylands is affected to some degree by various forms of land degradation.

There was an increase from 3475 million ha in 1984 to 3592 million ha in 1991, i.e., 117 million ha or 3.4%. The UNO Report forecasts, very rightly, that only 18 industrialised or oil-producing countries of 99 affected countries are believed to be able to combat the desertification of approximately 1.5 billion ha of their territories. For the 81 developing countries, whose 2.1 billion ha of lands are undergoing desertification, the problem needs external assistance. The analysis of soil degradation, in degrees, in areas of the world's drylands, indicates that major areas of degraded soils are confined to semi-arid and arid zones. Likewise, its analysis by types shows that the major soil degradation process in the dryland is wind erosion (512.4 m. ha) followed by water erosion (478.4 m. ha) then chemical (111.5 m. ha) and physical degradation (34.9 m. ha).

The analysis of soil degradation, by type and degrees, excluding hyper-arid zones, indicates that the major cause of soil degradation in these areas is water erosion (45.2%), followed by wind erosion (41.8%) then chemical (9.7%) and physical (3.4%), the dominant role being played by slight degradation (41.3%) and moderate degradation (45%), while strong degradation (12.6%) and degradation of extreme degrees being less significant (0.7%) only.

Three major factors responsible for soil degradation in drylands are: overgrazing (34.5%), deforestation (29.5%) and agriculture (28.1%). The percentage of the affected areas is largest in Africa (81%) followed by Asia (22%). It is evident from the figures 1,2 and 3 that the largest areas that are already experiencing the desertification hazard are found in Asia, Africa (deserts and territories with high levels of desertification) and Australia. The least effected areas are typical of Europe (moderate level) and North and Central America. A moderate level also exists in South America. Very high levels of desertification occur in Africa, Asia and South America primarily on territories occupied by developing countries. Land degradation in India is estimated to be 174 million ha already effected by soil erosion. Of this land 8.53 million ha was underlogged, 3.92 million ha was ravine and gutted land, 3.58 million ha was affected by alkalinity and 5.50 million ha belong to saline and sandy areas. However, since 1984-85 about 40 million ha are reported to have been recovered (Eighth Five Year Plan, p.33; 2.6.3).

Arbo Diallo, head of the interim Secretariat of the Convention to Combat Desertification said that " It directly affects more than 250 million and threatens a further billion world wide. In Africa, there are 400 million people living in arid areas who need help. Around 65% of African territory is made up of arid lands, a third of these deserts. In total 36 African countries are suffering from drought and soil degradation. The anti-desertification secretariat estimates losses in the region affected by degradation to be worth some $42 billion per year. In any African Sahel country the reduction of natural resources caused by desertification is equal to 20% of its GDP" (D+C, p.32, 1997).

Combating Desertification
The United Nations' Plan of Action, following the 1977 Conference as well as the Overview, describes measures designed

to actively combat desertification. Its working document was the "Plan of Action to Combat Desertification (PACD)", and the Governing Council of UNEP has regularly been convinced through its decisions in 1984, 1989, 1991 and 1992 that PACD is an appropriate instrument to assist governments in developing national programmes for arresting the process of desertification. Twenty-five major recommendations were made in 1977. Recommendations relating to an effective combining of industrialisation and urbanisation with the development of agriculture and their influence on the ecology of arid lands were also included, based on the experiences in the USSR.

The Earth Summit, (The United Nations Conference on Environment and Development) took place at Rio de Janeiro from 3 to 14 June 1992 which consisted of 27 principles (Yearbook UNO, 1992, pp. 670-681). Section II of Agenda 21 covered the conservation and management of resources. Three chapters dealt with managing fragile ecosystems. Agenda 21 called for strengthening the desertification knowledge base, developing information and monitoring systems for regions prone to desertification and drought, combating land degradation and promoting alternative livelihood systems in areas prone to desertification. In addition, the conference proposed integrating comprehensive anti-desertification programmes into national development and environmental plans, developing comprehensive drought preparedness and drought relief schemes for drought prone areas, designing programmes to cope with environmental refuges and promoting education on desertification control and management effects of drought. The conference called for the General Assembly to establish an Intergovernmental Negotiating Committee to elaborate an International Convention to Combat Desertification in countries experiencing serious drought and/or desertification. On 22 December 1992 the General Assembly adopted resolution 47/190 without vote. The Assembly would organise a regular review of the implementation of the Agenda. The Desertification Convention was finalised on 17th June 1994 in the meetings by the Intergovernmental Negotiating Committee for the Desertification Convention (INCD). It has a greater emphasis on a 'bottom up' approach with a greater focus on Africa. Many post-agreement negotiations have been made since its establishment. The tenth session (NICD 10) was held in New York from 6-17 January, 1997. Its negotiations resumed in Geneva from August 18-22, 1997.

The facts given in Chapter II of 1991 Report describe the roles of the UN System and the International Community, Regional and Sub-regional Cooperation and Actions at National Level. The implementation of the PACD during 1978-91 was evaluated several times and it can be concluded that there has been a failure to respond adequately to the requirements of the PACD, despite DSECON'S (Consultative Group for Desertification Control) activities, because of the apparent unwillingness on the part of the affected countries and the donors to make the Plan work as originally conceived. Between 1978 and 1985, some 50 projects costing US $ 15 million were completed and in 1985 there were some 20 projects under implementation at a cost of US $ 51 million. The following networks were established under UNEPC, e.g., Sand-Dune Fixation North Africa and Middle-East (ESCWA); Afforestation Latin America (ESLAC); Regional Network of Research and Training Centers for Desertification Control in Asia and Pacific (ESCAS/UNEP/UNESCO); NGO Network on Research and Information Development of Sustainable Livelihoods in the Arid and Semi-Arid lands in Africa (ECA); Watershed Management Network -- SADCC region of Africa (ECA); Chaco Arid Zones Network Argentina (ECLAC); Dendro-Energy Network-Peru (ECLAC). Other global level networks included the MAB National Committee of UNESCO, as well as the MAB International Network of Biosphere Reserves. These international collaborative projects also included projects on deserts, i.e., Sahara, Somali-Chalbi and Kalahari-Namibia. In all 529 projects were implemented by 1990 through various agencies. UNEP also sponsored 61 courses regarding training, seminars and workshops. Several anti-desertification programmes have been undertaken by FAO in 20 sub-Saharan countries the example being Keita project in Niger.

The subregional mechanisms include the fight against desertification in Sahel region by United Nations Sudano-Sahelian Office (UNSO), Committees in Dakar and Arab countries as well as PACD Action Plan regarding 'Savanisation' and Sahelisation', The Lagos Plan of Action, Cairo Programme all giving high priority to drought and desertification. Likewise Arab League Educational Cultural and Scientific Organisation (ALECSO) has ventured and cooperated the North African Green Belt Project. The Arab Center for Studies of Arid Zones and Drylands works in Western Asia.

So far only some 20 countries of the 99 affected have developed national programmes to combat desertification. The afforestation campaigns of Algeria, India and Kenya show massive public participation. Several eco-village projects have been launched in China and Africa.

Implementation of Recommendations

The specific recommendations of PACD numbering 28 are concerned with assessment of desertification, combination of industrialisation and urbanisation with the development of agriculture and their effects on ecology in arid lands; correlative anti-desertification measures; monitoring; socio-economic aspects; insurance against the risk and effects of drought; strengthening science and technology; establishment of national machineries; integration of anti-desertification programmes and international action. The Report concludes that the area of lands affected by desertification is not decreasing although some trees were planted throughout the world and some areas of shifting sands were stabilised.

Chapter III of the Report has parts A and B. Under A, policy guidelines deal with the general goal and practical targets. To reach these goals, six main environmental/developmental targets and 11 targets for supporting measures have been set for the year 2020. The 26 main principles in implementing PACD form the basis of the global anti-desertification strategy. These include six principles adapted from Den Bosh-Declaration (1991).

The course of action part B is concerned with (a) nationally adoptable guiding policies, and (b) practical steps. It is assumed that National Plans of Action to Combat Desertification (NPACD) should be fully integrated into national programmes of socio-economic development and accorded their proper places, priorities, and resources etc. The practical steps emphasise general priorities regarding preventive, corrective and rehabilitation measures.

The six steps to be taken under recommendation one are related to the introduction of improved landuse systems with an aim to introduce an integrated approach to the utilisation of every piece of land; to introduce land/water/crop management systems; to stabilise rainfed croplands; to introduce improved rangeland/husbandry management systems; to undertake major afforestation/reforestation programmes and to undertake, whenever appropriate, a major campaign on stabilisation of shifting lands etc. Other recommendations are set to develop and introduce appropriate and improved agricultural and pastoral technologies that are socially and environmentally feasible and compatible with new landuse systems; to establish adequate communication infrastructure and sufficient processing and marketing facilities; to develop and conserve available water resources and to introduce improved water management systems; to reclaim for productive use or to protect for natural rehabilitation, as appropriate, severely desertified lands. The recommendations regarding supporting measures are to establish or to strengthen the national institutional capabilities; to launch nation-wide major anti- desertification awareness/training campaigns; to introduce a 'loop model' in the existing or newly established extension service; to finalise the operative large scale local and national assessment of the current status of desertification; to develop, adopt through appropriate national legislation and introduce, institutionally, a new set of national environment/development oriented landuse policies and to develop and introduce effective national insurance schemes against recurrent drought and famine.

Regionally the eco-geographical regions of the world should be fully utilised, i.e., Mexico-USA; China-Mongolia-USSR, India-Pakistan, Afghanistan-Iran-USSR etc. Establishment of small sub-regional offices, analogous to UNSO needs proper consideration.

Internationally, participation and cooperation is needed in many areas like mobilisation of financial resources, trade policy, technical assistance, technologies, monitoring, information exchange and legislation. Over every five years, beginning in 1995, UNEP and UNDP should jointly review the implementation of the PACD and programmes of Rio-Declaration of 1992 (UNO, UNEP Report,1992).

Financial Constraints

The executive summary of the Report (UNO, UNEP Report, 1992) states that the cost of meeting the minimum objectives of halting the spread of desertification, i.e., the cost of urgent direct preventive measures in non-affected but vulnerable or only slightly affected irrigated lands (70% of their total area); rainfed croplands (53% of their total area); and rangelands (27% of their total area) amounts to about US $ 1.4 to 4.2 billion a year for the 20 year programme. This should be complemented by the cost of direct corrective measures in moderately affected irrigated lands (23% of their total area); rainfed croplands (40% of their total area) and rangelands (28% of their total area) amounting to nearly US $ 2.4 to 7.2 billion a year for the same 20 year programme. Out of the total sum of US $ 3.8 to 11.4 billion a year, US $ 2.2 to 6.6 billion a year is needed for financing the actions in 81 developing countries, half of this sum could, at least, be raised by the countries themselves.

The World Bank's 1995 environment annual report described the details of some US $ 5.6 billion investment out of the $ 10 billion sanctioned as loans for 137 projects in 62 countries (*Hindustan Times*, New Delhi, Dec. 13, 1995). The following lines present the sorry state of affairs regarding our efforts. A special session was called at New York in June 1997. Ismail Razali, Malaysia's ambassador to the UN, who presided over the summit, opened by describing the actions taken since Rio meet as "paltry". He said "There was a recession of spirit" afflicting the signatories to Agenda 21, "Damning statistics show that we are heading further away, and not towards, sustainable development." However, some achievements have been made in setting dates and percentage targets for reducing emissions of greenhouse gases in the meeting held at Kyoto, Japan, in December, 1997 (*Nature* 26th June 1997, p.836).

Need of Future Steps

(1) The United Nations Convention to Combat Desertification is a framework which requires implementation by involving cross-section of groups, communities, institutions and governments.

(2) The United Nations Framework Convention on Climate Change (UNFCCC), the Convention to Combat Desertification (CCD) and the Convention on Biological Diversity (CBD) have a great deal in common. These, having shared problems, need closer collaboration not only among the secretaries of the Conventions but also among their political processes at both the national and international levels.

(3) The Association of Small Island States (AOSIS) and developed countries need each other's support in addressing their respective priorities of climate change and desertification.

(4) Countries need to develop integrated approaches to the Rio Conventions forging closer links among ministries involved.

(5) In Article 4 of the Convention, general obligations for all Parties are outlined. All Parties are expected to implement these obligations individually or jointly through existing or prospective bilateral and multilateral arrangements or a combination of these. Developing Country Parties should be assisted in this process.

(6) It is time that African Country Parties, Affected Country Parties and Developed country Parties lived up to their commitments in order to give new impetus to the implementation of the convention (Otinda, 1997, pp. 12-13).

(7) There is need of a Global Participation Fund self-administered by developing countries, to support their active involvement in global negotiations (Kaul, Grunberg and Stern, p.1999).

(8) As UNDP Administrator James Gustave Speth notes appropriately: International cooperation today must remain focused on aid on the poorest countries. But it must also go beyond aid----the public good agenda is a new, additional challenge (Kaul, et al.).

References

Darkoh, M.B.K. (1989), 'Combating Desertification in the Southern African Regions: An Updated Regional Assessment', UNEP, Desertification Control Programme Activity Center, Moscow.

Diallo Arbo, 1997, 'Call to Fight Desertification' in 'Development and Cooperation' No.4, July/August, Frankfurt, Germany. p.32

Drenge, H., Kassas, M. and Rozanov, B. (1991), 'A New Assessment of the World Status of Desertification', Desertification Control Bulletin 20, pp.6-18.

Eighth Five Year Plan (India), p.33; 2.6.3

Gerasimov, I.P. (1983), 'Geography and Ecology', Progress Publishers, Moscow, p.129.

Grainger, A. (1990), 'The Threatening Desert: Controlling Desertification', Earthscan/UNEP, London/Nairobi.

Kaul Inge, Grunberg Isabelle and Stern, Marc A. (eds) (1999), 'Global Public Goods: International Cooperation in the 21st Century. Publication for United Nations Development Programme (UNDP) by Oxford University Press, in D+C, 4,1999, Frankfurt, Germany, p.35.

Otinda, Peter A. (1997), 'The Desertification Convention Parties Should Honour their Commitments', CAN Newsletter, No.21, pp.12-13.

Report of the Executive Director on Status of Desertification and Implementation of the United Nations Plan of Action to Combat Desertification, Governing Council, Third Special Session, Nairobi 3-5, February 1992, UNEP/GCSS.III/3,15 October 1991, pp.3, 16, XII. See also Williams, M. McCarthy and G. Pickup in Australian Geographer 26 (1) May, 1995 tables 1 and 2 p.27-29 cited below for causes, consequences and impediments.

Scoging, H. (1993), 'The Assessment of Desertification', Geography, 78 (2), pp.190-193.

Williams, M.A.J. and Balling, R.C. (1994), 'Interactions of Desertification and Climate', WMO and UNEP.

Williams M., McCarthy, M. and G. Pickup et al. (1995), 'Desertification, Drought and Landscape: Australia's Role in an International Convention to Combat Desertification', Australian Geographer, 26(1), May, p.23.

Yearbook of the United Nations, 1992 Vol.46 Dept. of Public Information, United Nations, New York, Martinuss Hijhoff Publishers, Dordrecht/Boston/London, pp.670-681.

4

Water Resources Changes and Land Degradation in the Valley of Lakes, Mongolia

N. Batnasan

The Valley of Lakes is located in the northwest of Gobi Desert region of Mongolia and borders with Gobi-Altai Mountains in the south Great Lakes Deposition in the west and Khangai Mountains in the north. This region consists of four main river basins Baidrag, Taats, Tyi and Ongi, which originate from Khangai Mountain and end in closed lakes, such as Boon-tsagaan, Taatsiin-tsagaan, Orog and Ulaan, respectively. Total area of this region is 99,797 km². (Figure 1). The climate is harsh with extremes of heat in summer and cold in winter. Availability of water and other renewable natural resources is relatively limited and natural ecosystems are fragile, easy to degrade and slow to recover.

In generally, there are more than 3500 lakes, 3800 rivers and hundreds of other water bodies in four main geographical regions of Mongolia: Altai, Khangai and Khentei Mountain, Eastern Steppe and Gobi arid region. In spite of the aridity of the Gobi region it has a concentration (36.5%) of lakes. Total water resources of Gobi's lakes calculated as 118.8 cubic kms, from which only 3.84% (4.565 cu. kms) corresponds to the Valley of Lakes Basin. Most of lakes are salty (1.75-3.86 g/l) and some

Figure 1
Basin of the Valley of Lakes, Mongolia

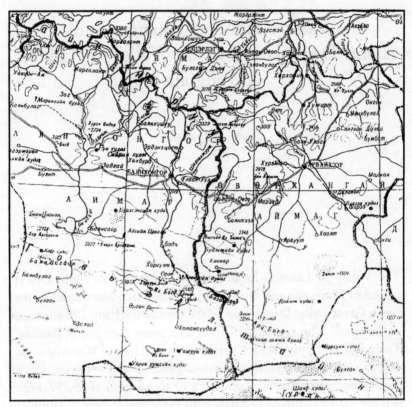

temporary large lake's mineralisation is reached up to (338.1 g/l) lake Ulaan. Annual mean of river discharges varies from 0.28 to 15.5 cubic m/s.

The rivers and lakes are becoming important natural resources for regional development and economic and social life of this region. Since water resources in such arid region is very sensitive to climate changes, mismanagement of natural resources in dry years, one could expect great loses, which occurred several times in previous years, causing huge degradation of land cover.

Therefore, discussed here are the main features of the morphology of lakes, water resources changes, landuse practices, irrigation activities, land degradation problems and possible solutions for effective use of environmental resources.

Morphology and Water Resources Variability

Main morphological feature of the closed lakes characterised by their relatively less depths, big areas and variation in time (Table 1). It means that lake shorelines decrease with small fluctuations in water level due to the very flat profile of the lake beds, which in turn increases evaporation from the lake surface. Open water evaporation from these lakes is estimated to be 920-1000 mm/year (N. Batnasan et al., 1987, 1988).

Table 1

Morphological Characteristics of Lakes in the Valley of Lakes, Mongolia
(J. Tserensodnom, 1971 and N. Batnasan, 1998)

Lake Names	Altitude (m)	Area (sq. km.)	Length (km)	Width (km)	Max. Depth (m)	Volume (Cu km)	Perimeter (km)	Catchment Area (km)	Mineralisation (g/l)
Boonitsazaan	1312	252	24	19	16	2.355	81	33500	3.86
Orog	1217	140	31.8	7.7	5.0	0.42	75	10500	1.75
Adgiin-tsagaan	1285	11.5	7.6	2.2	1.5	0.009	22.2		5.87
Taatsiin-tsagaan	1234	9.9	4.1	3.2	2.3	0.01	11.8		5.01
Ulaan	1008	175							338.1

Hydrological condition of rivers, inflowing into the closed lakes are mainly characterised by the following situation. River discharges increases along its length until it reaches the pediment zone, where its run-off reduced by 2-5 times and only small amounts of water reaches the lakes. In dry years, it is often recorded that these river flows do not reach the lakes, infiltrating or evaporating on the way.

Water resources of this region varies in different seasons and long-term time series. For analyses, some monitoring results from

hydro-meteorological stations in the lake Boon-tsagaan, rivers
Baidrag, Tyi and some other stations, located in the basin have
been used (Source: Institute of Hydrology and Meteorology, 2000).
The results shows that, there are similarities in seasonal
distributions of precipitation in the basin and river discharges and
lake level changes. Most precipitation occurs from June to August
(about 70%) and highest river discharges in July to August and
maximum lake levels are reached in August and September. It
indicates that, formation of river water resources largely depend on
the amount of precipitation in the mountains and in turn it
considerably affects closed lakes level changes (Figure 2).

Figure 2

*Seasonal Variation of Precipitation (P in Galuut), Baidrag River Discharge
(Q-Bayanbyrd) and Boon-tsagaan lake level (H)*

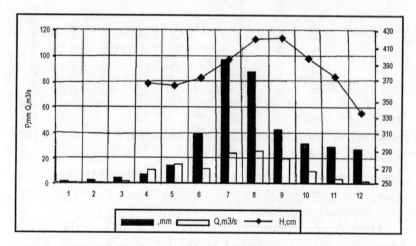

As investigation results shows that, the lake morphological
characteristics of this basin are quite changeable, because of
considerable dependence on water resources of tributary rivers.
Lake Boon-tsagaan is the largest in the Valley of Lakes with total
area 252 sq. km in 1971, 217.5 sq. km in 1989 and 338.8 sq. km in
1992. During the end of 1980s the depth of this lake was reduced
by 2 meters since 1971, while others (Orog, Taatsiin-tsagaan and

Ulaan) have dried up. It is a crucial problem for the environment and social life of the region.

In earlier periods, variation in the climatic and hydrologic conditions in the Valley of Lakes Basin were quite large, due to which information about lake level changes were recorded. For example, in 1936, the Orog lake was dried up and in 1952-1953 it was splitted into two parts and in 1967 the size of lake Boon-tsagaan was increased. The Ulaan is very shallow (1-2 m), but has a very big area and its water regime largely depends on climate. Therefore, it is a most dynamic large lake in this region.

However, the environmental problems caused by the drying up these lakes in the same time, which happened at the end of 1980s were never recorded before (Figure 3). Certainly, there were a number of reasons for it, firstly, it was clear that, the dominant factor was climate change for it, secondly, human activities mainly on the river banks and around the lakes played quite a large role in the degradation of the environmental conditions and water resources.

Climatic Factors

The main climatic factor for water resources changes is precipitation. Annual precipitation near the lakes is less than 100 mm, and in the Khangai Mountain it reached up to 300-450 mm. The long-term mean of precipitation in this basin is 218 mm and its maximum occurred in 1964, 1976 and 1994 (Figure 4). From end of 1970s to the end of 1980s, it was below the long-term average. During this time, many small tributaries of the main rivers dried up and ground water level fell by 2-3 meters and in some places even more.

Human Activity Factor

Main water use in this basin is for irrigation. In spite of that, the total area of irrigated lands in the Gobian region is only about 1,800 ha, the consumption of river water is always excessive because of lack of interest in conservation. For instance, because of

Figure 3
Change of Surface Area of Some Lakes in Valley of Lakes
1. Boon-tsagaan, 2. Orog, (1949, 1984, 1989)

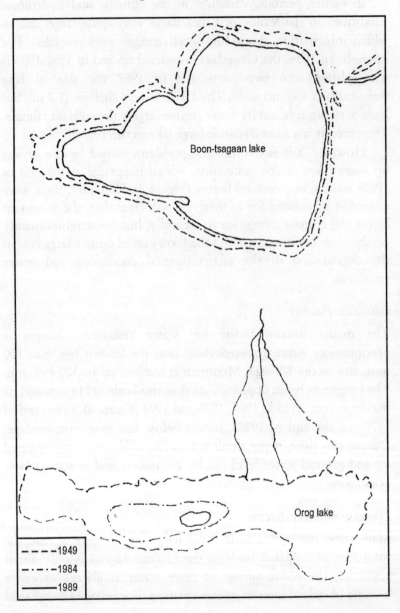

Boon-tsagaan lake

Orog lake

- - - - 1949
- · - · 1984
——— 1989

Figure 4

Long-Term Changes of Precipitation, River Discharges and Lake Level
5-precipiation in Bayankhongor and 6-Galuut, Q1-river discharges Baidrag
and Q2-Tyi, H1-Boon-tsagaan lake level

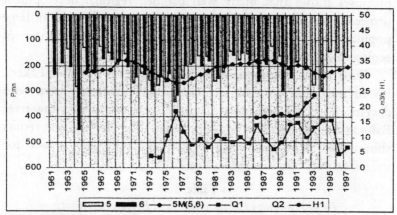

leakage from irrigation canals the efficiency of the irrigation is only 30-45% in most irrigation systems in the river basins of Baidrag, Tyi and others. In dry years, total amount of water for irrigation from river Tyi reached up to 40% of annual run-off, which is considerably affected by the reduction or drying up of the Orog Lake.

Land, Water Use and Degradation

Rivers, lakes and ground waters are the main source of water for residents in the Valley of Lakes basin. If there is a lake, people living nearby use it as a water supply, especially for livestock watering. The river waters are mainly used for irrigation purposes, increasingly, reducing runoff from inflowing rivers in the closed terminal lakes. Thus, development of irrigation in lakes basins as well as the reduction and drying up of Lakes lead to serious environmental and socio-economic problems.

Human activities in many water resource areas are recognised to be an increasing threat both to water quantity and quality and to the environment of these resources. The demand for irrigation schemes and uncontrolled cutting of forests in the upper watershed areas are causing reduction of river flows, lake levels and soil moisture content. The increasing use of water for irrigation purposes shows that many well planned regional development projects have adverse affects, including forced settlements of large populations, overgrazing, eutrophication and degradation of ecological systems.

Despite the shortage of water resources in the Valley of Lakes a lot of irrigation systems and canals for plantation have been developed. There are more than 10 such systems in the river Baidrag and about 6 in river Tyi's basin. Our investigation results shows that, in each year up to 40% of river water is used for irrigation purposes in these basins (Table 2).

Table 2
River Hydrology Characteristics and Irrigation Activity

River and lake names	Catchment area (sq. km)	River length (km)	Discharges (m3/s)		Number of irrigation systems	Water and for irrigation (m3/s)
			In the middle of rivers	Near the lakes		
R. Baidrag and Boon-tsagaan	27277	310	7.4	0.48	>10	1.80 (24.3%)
R. Tyi - I. Orog	9410	243	1.91	0.61	~6	0.56 (36.8%)
R. Ongi-I. Ulaan	52920	435	1.52	0.0-0.5		
R. Taats-I. Taatsiin-tsagaan	9190	200	1.12	0.0-0.6		

Development of different kinds of agricultural activities in the river basins led to more pressure on land through; landuse changes and increasing cattle density. This in turn effects overgrazing, trampling, soil erosion and sand movements. Overgrazing is very often observed in pastures as a result of large herds of livestock continuously staying near river valleys and shorelines of lakes. In the Gobian Lakes basins more than 65,400 ha of pasture are overgrazed each year. More than 200 ha of grass cover has been cleared for plantations of medical plants near Orog lake in 1989, several hectares of land have been covered by sand dunes and about 2 metres of sand accumulated on the Orog lake bed when it dried up in the end of 1980s. These results indicate that, the role of vegetation and soil cover are quite important in environmental stability in arid region and it requires very careful management techniques and supporting measures to sustain the integrity of such a fragile system.

It is very common that, opinions of the local people living near the lakes and upper part of river valley are quite different with respect to the development of irrigation activities within the basin. People near the lakes often were against the activities carried out in this basin, because they blamed the irrigation schemes operated in the upper part of rivers for reducing the lakes level and causing it to dry up. These people insist on stopping further operations for irrigation, but others do not. It means that, the traditional water use and environmental resource management approaches no longer provide sustainability and new integrated approaches are needed to reconcile conflicting interests on how to use and conserve environmental resources.

Conclusion

The Valley of Lakes basin is a unique region for Mongolia, representing mountain, steppe, arid land and wetlands. It should be protected in its complexity and developed in an environmentally friendly way.

The closed lakes in this basin are becoming very clear indicators of climate changes. Historically, the hydrological regime of theses lakes were naturally changed, but during the last few years, human activities are greatly influencing it, resulting in considerably reducing river flow and lake levels on the background affect of climate change. Therefore, water use should be accepted as a component of the environment and must be an integral part of the total environmental management package. Through water resources management policy, landuse and human activities can be guided and controlled. Thus, we should put more emphasis on very careful management of water resources and landuse changes in the Valley of Lakes and their conservation.

References

Batnasan N, Tserensodnom J, (1987), Water thermal regime of lake Boon-tsagaan, //Geographical problems of Mongolia (GPM), No. 27, Ulaanbaatar, (in Mongolian), pp. 18-24.

Batnasan N, Sevasbanov D.V, (1991), Humidity changes and lake conditions in Mongolia, //Geography and natural resources, No.2, Novosibirsk, pp. 177-183 (in Russian).

Batnasan N, (1998), Hydrological systems, water regime and natural development of Greet Lakes in Gobi Desert region, Ph.D. Dissertation in Geographical Science, Ulaanbaatar, p.263

Batnasan N, (1999) Morphology and long-term water level fluctuation of lake Uvs, in the booklet of International Conference: Global Change and Lake Uvs, Ulaangom, Mongolia, 1999, p.94-108

Batnasan N, (2000), Hydrological systems of Gobi Desert's Great Lakes Basin and their Environmental Responses to Climate Change, In Abstract volume of the 29th International Geographical Congress, 14-18 August 2000, Seoul, Korea.

5

Land and Land Degradation: Conceptual Overview

N.K. De and Ananya Taraphder

The common meaning of land is the solid portion of the surface of the globe. It is a product of nature, a three dimensional dynamic body, on which human civilisation has developed and sustained. It is that part of the earth which supports life and has multidimensional aspects and hence its necessity as well as utility is immense. It is also the carrier and supplier of the day-to-day needs of the society. With the increasing demand of the society, multifunctional use of land begins. Since the land area is fixed, so a piece land is being used again and again. As a consequence, land degradation takes place.

Degradation is a process of wearing down of the land surface through natural agents. It may be by fluvial action, glacial action, aeolian action, marine action or by other agents. It involves the degeneration of land surface. Land degradation is mainly caused by water. It considers physical as well as chemical leaching and loss of soil from land surface. It is the deterioration of natural properties of land. Land degradation is a serious problem all over the earth surface.

Concept of Land

The concept of land varies with time and individuals. In the early phase of human civilisation the vast land along with the diversified physical phenomena was merely treated as the aesthetic one. But with time the value of land appreciated progressively. The increasing pressure of population on land has compelled the geographers and others to think about land afresh.

Scientists belonging to different disciplines have considered the land from their own viewpoints to find out its versatile aspects, ultimately to meet the increasing demands of people. To a geographer, it is treated in the spatial context, to a geologist, land is a complex evolutionary product. To the geomorphologist, land is a tract with bare, steep/gentle slope and/or is merely an erosional/depositional product. To the pedologists, the land is a product of nature evolved through the combined activities of parent material, climate, vegetation as well as organisms and process through time. To the pedogeomorphologist, land and soil is inseparable and he considers land as an integrated outcome of pedologic and geomorphic processes. To an edaphologist it is looked as the habitat for plants.

To a botanist, the land is the space for growing plants. An agronomist looks at it as the field for agricultural practices. To the eye of an economist, the land is a valuable product and it is measured in terms of money. To the planner, the land is a blanksheet of paper without human interactions where agriculture can be practiced, city/town can be established, roads can be constructed and so on, as per the suitability of land and facilities provided. A house owner has different conception of land, a housewife appreciates the land as the space for kitchen garden as well as for the decoration of home. To the artists, the land has romantic and aesthetic value.

Thus, it may be generalised that land has multifarious aspects; (i) as space, (ii) as situation, (iii) as property, (iv) as capital, (v) as factor of production, (vi) as a location, (vii) as a resource, (viii) as matter of conflict and so on. However, the modern concept of land is something more than what is stated above. A part of the land surface that one can see in nature is not a simple but a complex one.

Land Complex

Land is a three dimensional dynamic body with complex nature and may be compared with other living beings. If one behaves well with land it reciprocates positively, if one behaves badly, land responses negatively, if one is to understand the nature, behaviour and quality of a land surface, it is to be treated very carefully and tenderly for its sustainable utilisation.

Land complex may be defined as an area or group of areas, throughout which there is a recurring pattern of topography, soils, vegetation etc. It consists of landform assemblages with its associated pattern of soil and vegetation and it can be sub-divided into distinctive land units of which they are composed of.

The concept of land complex visualises that each part of land surface is the end product of an evolution governed by parent geological material, geomorphological processes, past and present climate and time. The land surface has thus been shaped to its existing land units, each developing in the process its own hydrological features, soil mantle, vegetation communities, animal population and a range of micro-environments. It is further emphasised that land units maintain their character as a recurring topographical units together with characteristics of soil, vegetation etc. Thus the concept of land complex is a scientific one as it is based on topography, soil and vegetation, correlated with geology, geomorphology and climate.

Land Degradation

Broadly speaking land degradation may be of two types, physical degradation and chemical degradation. These two types of degradation mostly occur simultaneously. Chemical degradation is generally confined to humid areas having good amount of rainfall. This occurs in two ways: (a) chemical matters are dissolved and transported with surface running water and (b) the leaching of chemical matters and nutrients down the soil profile, thus creating the upper land/soil which is infertile and changing the existing land/soil quality.

Physical degradation is common almost everywhere whether humid or arid/semi-arid. Land/soil is eroded as well as displaced by the action of natural agents in the form of surface soil. Sometimes negligible amount of finer soil or colloids also move down the soil profile and are deposited within the profile. It is cemented with an admixture of different chemical matters by physical degradation processes. There is however, a major difference between surface removal of soils and the degradation of soils through profiles. The soil particles that are washed away or blown away through the land surface are transported far away and are deposited elsewhere, whereas the soil particles/colloids/ chemical matters that are leached through the profile cannot be transported long and are deposited at a particular layer within the profile. Thus the surface removal of soil causes great havoc to the human civilisation mainly.

Man is struggling for his existence on this earth. Exponential increase of population with limited land and water resources has created grave concern in the world today, the environment is being threatened by the reckless activities of man. Man inherits the world of soils, water, air, plants and animals and he creates the world of social, economic and political institutions.

Land degradation occurs both by natural as well as by human agents. However the problem of land degradation as a result of man's activities is alarming. Natural resources, particularly land, soil, air and water are seriously affected. Though technology improves the quality of life, but it lowers down environmental conditions in many cases. Fertilizers and pesticides are important factors in increasing the crop production, but eventually they threaten to create an environmental imbalance. Due to the pressure of population man is compelled to use land very intensively and employ modern technology more and more to meet the increasing demand. This has created numerous problems in regard to land degradation.

There are demands for land for activities like agriculture, forestry, grasslands, urban and industrial development, transport, mining etc. All these activities have harmful effects on land. The main source of population in rural areas is the residue from fertilizers, pesticides, dirty water, household wastes etc., while in urban areas garbage, household wastes, industrial wastes, papers, cans etc. are responsible. In general, the inadequate sewerage system, household wastes and slums are responsible for polluting land and soil, of both rural and urban areas.

Factors and Types of Land Degradation

Factors responsible for land degradation include climate, concentration of rainfall and effects of seasons, parent material, high water-table, topographic variation and slope of land surface, carrying capacity of running water and floods, nature of soil and plant cover, crop rotation and cropping pattern, over grazing, etc.

The types of land/soil degradation are: (i) water erosion – stream bank erosion, sea-shore erosion, gully erosion, rill erosion, sheet erosion, slip erosion, splash erosion, (ii) wind erosion, (iii) glacial erosion, (iv) damage due to excess of salts, (v) chemical degradation, (vi) biological degradation and (vii) fertility erosion. Though these are active almost everywhere on the earth surface, they are very active in the tropical countries. Since it is difficult to quantify chemical and biological degradation, the study has so far been confined in the tropical countries to water erosion, wind erosion and damage caused due to excess of salts mainly. The effects of chemical, biological and fertility erosion, however, are felt strongly but their exact quantification has not been made so far in these areas.

The problems of land degradation are different from place to place and from region to region to region depending on (a) natural and (b) man's activities on the earth/land surface which may be grouped into three categories according to the degree of seriousness as follows:

Category-I

This category includes four land degradation cases such as (i) soil erosion, (ii) salinity and alkalinity, (iii) organic wastes, and (iv) infectious organisms.

(i) *Soil erosion*: Soil erosion is the most serious factor for land degradation. It removes huge amount of nutrients in suspension or in solution from place to place causing depletion or enrichment of nutrients. Tremendous amount of sediments are removed by water and wind. Besides the loss of nutrients due to surface erosion, degradation through rills and gullies makes the land unsuitable for crop cultivation. Landslides in the hilly tracts and subsidence of land affect the high ways, dams and habitations. The loss of nutrients particularly nitrates and phosphates due to soil erosion is the major factor of water pollution. Measures for soil conservation will go a long way in improving environment and preventing degradation of soil.

(ii) *Salinity and alkalinity*: In arid and semi-arid areas salts and alkali accumulate and threaten the agricultural production. Besides reduction of agricultural production, accumulation of salts and alkali results in the concentration of toxic elements which are responsible for health hazards both for men and animals. The problem of salinity and alkalinity can be solved by adopting controlled irrigation system and by rainwashing to restore normal crop production.

(iii) *Organic wastes*: In urban areas most of the land degradation problems are due to the high concentration of population. Municipal and industrial wastes contain detergents, carbonates, phosphates and many other salts that are responsible for land degradation. These wastes which are rich sources of soil fertility, pollute the land surface by discharge of heavy metals, DDT and different pesticides.

(iv) *Agricultural wastes*: Agricultural wastes consisting of animal wastes, crop wastes and agro-industrial wastes which in many ways are responsible in polluting the rural environment create an unhygienic atmosphere. It is necessary to develop a system of recycling of nutrients from these wastes. Use of a mixture of organic manures and inorganic fertilizers should be introduced into the farming system.

(v) *Infectious Organisms*: Infectious organisms, pests and various soil borne disease are responsible for land/soil degradation. It is necessary to control infectious organisms, pests, and diseases by spraying insecticides and fungicides and by introducing disease resistant crops. However, it is better to use more disease resistant crops than to spray pesticides and fungicides which not only pollute soils but also cause health hazards.

Category-II

Land degradation problems in this category comprise (i) radio-activity, (ii) industrial wastes, (iii) pesticides, and (iv) heavy metals.

(i) *Radio activity*: Radio-activity materials react with soil in the same way as the ordinary elements do. These contaminants normally result from mines and refineries of uranium, thorium power plants, nuclear-reactors and wastes from medical and different scientific laboratories. It is also caused by various minerals and rocks. The decontamination of soil can be done by continuous cropping, removal of surface soil (about four inches in depth), application of gypsum etc.

(ii) *Industrial wastes*: Industrial wastes which cause atmospheric pollution and soil degradation are due to the release of stack gases (fly ash, sulphur dioxide, fluorides etc.) and other industrial and mining wastes (acidic effluent from iron, zinc, coal, copper, aluminium and so on). All these have a disastrous effect on land/soil, water and crop, causing health hazards of both man and animal.

(iii) *Pesticides*: Pesticides have had a technological boost in modern agricultural production. But it has been observed that several pesticides which persist in soil for a long period show undesireable effect in crops and microbial activity in the soil body. As such the use of pesticides must be minimised.

(iv) *Heavy metals*: The concentration of heavy metals like lead, zinc, cadmium and arsenic in plant and animal tissues are responsible for toxicities in human beings. These heavy metals also concentrate more in surface soil than in subsoils. Preventive measures should to be taken to eradicate these pollutants.

Category-III

Land degradation problems of this category are of less significance. They are due to fertilizers and detergents. At the present moment though they are not so hazardous, careful attention needs to be paid on this problem.

(i) *Fertilizers*: The main concern of the application of fertilizers in the agricultural field is the washing out of nitrogen, phosphorus and potassium into inland water bodies because of surface run off and soil erosion. The nitrification of inland water bodies, lakes, rivers etc. caused by the accumulation of nitrates, phosphates and molybdates from NPK fertilizers appears to be of great concern.

(ii) *Detergents*: Detergents have little effect on degradation of land/soil. In and around towns and cities, the amount of detergents used is usually very high as compared to rural areas. The polluting effect of detergents on land/soil through sewerage is accordingly high. Care should be taken in this regard, and its excessive use avoided.

Concluding Remarks

From the above discussion it is quite evident that land degradation is a major problem which the society is facing today all over the World. The types of land degradation and the associated problems may be different in respective environmental situations both at macro and micro levels. The strategies also varies accordingly from country to country. The present paper is a humble attempt to go into the insight of the concept of land, land degradation and the resulting problems.

References

Azzi, G. (1995) *Agricultural Ecology*. Constable and Co., London.

Bakker, J.P. (1952) A remarkable new geomorphological law, Proc. of the Konin Klijke Netherlandsche Akademic Van Watenschappen, Sr. B., Parts I, II and III.

Christian, C.S. (1958) The concept of land units and land systems. Proc. of the Ninth Pacific Sci. Cong., *Humid Tropics*. Vol. 20.

Cotton, C.A. (1948) Landscape as developed by the processes of normal erosion. Second edition, Christchurch, New York. pp.509.

C.S.I.R.O. (1963) Australian Land Research Series 7.

De, N.K. (1984) *Measuring Land Potentials in Developing Countries.* The University of Burdwan, Burdwan, West Bengal, India.

F.A.O./U.N. (1952) Land utilization in tropical areas. F.A.O., Development Paper, pp.10.

Galon, R. (1964) Hydrological research for the needs of the regional economy. *Problems of Applied Geography*, Warsaw.

Graham, E.H. (1946), *Natural Principles of Landuse*. Oxford University Press, London, New York, Toronto.

Mohr, E.C.J. et. al. (1954) *Tropical Soils*. Inter Science Publishers Ltd. London.

SMIC (1971) Report of the study of man's impact on climate, MIT Press, Cambridge, Mass. and London.

Strahler A.N. and Strahler A.H. (1976) *Geography and Man's Environment*, John Wiley, New York.

U.S.D.A. (1969) A manual on conservation of soil and water, Handbook for Professional Agri. Workers. Second Indian Reprint, pp. 208.

Wilkinson, H. (1963) Man and natural environment, Department of Geography, University of Hull, Occasional papers in Geography, No.1.

Zobler, L. (1962) An economic historical view of natural resource use and conservation, *Economic Geography*, Vol. 38. pp. 189-194.

6

The Changing Landuse/Land Cover Scenario in Nepal

Motilal Ghimire and Narendra Raj Khanal

Nepal with an area of 147,181 sq. km presents a great physical and cultural diversity due to its geology, altitude, relief and location. It consists of many landscapes of different geological formations. The Terai in the south is the extension of Indo-gangetic plain, the Siwalik is composed of tertiary sandstone, siltstone and shale; the Middle Mountains and the High Mountains are composed of phyllites, quartzite and schist and Tethys sediments. The altitude ranges from only 60 m in the south to 8,848 m in the north within a horizontal distance of 160 km. Because of this altitudinal variations, the climate ranges from tropical in the south to arctic in the High Himalayas in the north. The orientation of mountain ranges has caused for several micro-climatic conditions. Besides, Nepal lies on the cross road of a number of floristic regions including the Sino-Japanese, Irano-Turanian, Central Asiatic and Indo-Malayan. Because of this diverse bio-physical condition in the country, it presents more than 35 forest ecosystems and over 5,400 species of vascular plants including 240 species of endemic plants and 700 species of medicinal plants; 130 species of mammals and 800 species of birds (Gorkhali, 1991).

The total population of the country has more than doubled from 8.3 million in 1952-54 to 18.5 million in 1991 with an annual growth rate of more than 2% per annum. Annual population growth between 1981-91 was 2.1% (CBS, 1996). More than 90% of the total population is engaged in subsistence agriculture. Agricultural density in the country in 1991 was 807 persons/km^2. It is very high in High Mountain areas (1248 persons/km^2) and Hills (982 persons/km^2) than in Terai (664 persons/km^2).

Livestock are an integral part of Nepalese economy. Livestock population per head of population and per area of arable land is high by Asian standards. Animal population increased from 13.32 million in 1981-82 to 16.68 million with an annual growth rate of 2.52%. In terms of livestock unit, it has increased by more than 2% from 8.31 million in 1981-82 to 9.98 million in 1991-92 (CBS, 1996).

Fuelwood is the major source of energy. Fuelwood provides 83% of the total energy consumption in the country. Commercial sources provides only 6% and remaining 11% is obtained from agricultural and animal wastes.

In addition, fuelwood and timber, fodder and litter are collected from the forest which are then used for livestock feeding and agricultural manuring. Livestock are also grazed on forest land. The sustainability of the economy and ecology of the country depends on the balanced growth of crop-livestock-forestry sectors in the country.

Because of over dependency of primary resources and increasing demand for food, wood and fodder due to rapid growth in human and livestock population, infrastructure and commercial activities, without developing other alternatives, there is high competition among different landuses particularly among agriculture, grazing and forest. It is believed that the imbalance in these sectors has been increasing and this has been leading to a vicious cycle of poverty and environmental degradation (Ives and Messerli, 1989).

The present paper aims to discuss land cover/landuse in Nepal and its dynamics. Information on landuse/land cover in the country reported by various authors from different parts of the country in different periods has been collected, compiled and analysed in order to understand the causes, processes and rate of land cover/landuse change in the country.

Landuse/Land Cover Pattern

Landuse/land cover information for the whole country was published in 1986 by Land Resource Mapping Project (LRMP) based on air photographs taken in 1978-79 for the first time. Though major land cover types (cultivated land, reserved protected forest, wooded area, grass and scrub land and water bodies) in the form of toposheets based on vertical photograph taken from 1952 to 1958 and field verification were prepared for the whole country by the Surveyor General of India. But tabular data based upon this survey is not available yet. Cadastral survey has not yet been completed in the country. Moreover, the districts where cadastral survey is completed does not show the real picture of land cover/landuse because of the purpose of the survey itself and definition used in classifying land and legal provision were made for the ownership right of the land. National Sample Census of Agriculture is undertaken in every ten years (1961-62, 1971-72, 1981-82 and 1991-92) but comparison on the change in agricultural holdings beyond national and regional level is difficult due to frequent change in district boundary.

Landuse survey of 1978-79 indicates that agriculture land comprised of about 20% of the total area of the country (Table 1). Similarly, forest, shrub and grassland comprised of 38%, 5% and 12% of the total area of the country respectively. About 7% land has been defined as non-cultivated within agricultural land which includes forest, shrub and grassland. About 19% of the total area is occupied by snow, ice, glaciers, landslides, rock and water bodies. As far as the geology is concerned, steepness of the slope and climatic condition differs from one physiographic region to another, landuse/land cover pattern also differs. Nearly 60% of the total area of the Terai is under agricultural use followed by forest (28%). Similarly, 77% of the total area in the Siwalik is under forest cover followed by agricultural use. Nearly 40% of the total area in the Middle Mountain is covered by forests followed by agricultural use. Major land cover types in the high Mountains are forests (55%) and grassland (17%). More than 67% of the total area in the High Himal is covered by snow, ice and glaciers followed by grassland (26%).

Landuse/Land Cover Change

Agriculture

Estimated landuse/land cover data for 1985-86 by physiographic regions is presented in (Table 1). It is estimated that agricultural land has been increasing annually by 0.40%. The highest growth in agricultural land is estimated in Terai and Siwalik region than in the Hills and Mountains. Negative growth is reported from High Mountain areas. Case studies from different parts of the country in different periods indicate an increasing trend in cultivated land, except in Bamti-Bhandara in the High Mountains and Dhankuta area in the Middle Mountain region (Table 2). Recent study on the change in land cover between 1978-79 and 1992 in the hills of Eastern and Central Development Regions shows an annual increase in agricultural land by 0.2 and 0.48% respectively. Study in Upper Pokhara and Nakhu watershed (Middle Mountain) indicate an increase in agricultural land by more than 0.7% per annum. It is also reported from Upper Pokhara that the rate of increase in agricultural land was higher in mid 70's than in the 80's. Another study in Jhikhukhola indicates a rapid increase in agricultural land at present than in the past. The rate of increase in agricultural land in Chakrimakri and Chure watersheds in the Siwaliks was reported to be more than 69% per annum between 1954 and 1979.

National Sample Census of Agriculture shows that total area of all holdings increased by 54% over the last thirty years from 1.69 million ha in 1961-62 to 2.6 million ha in 1991-92 (CBS, 1994). Arable land (area under temporary crops) in the country increased from 1.592 million ha in 1961-62 to 2.3234 million ha in 1991-92 with an annual rate of change by -0.16% between 1961-62 and

Table 1
Landuse/Landcover Types and Change Between 1978-79, 1985-86 (%)

	Cultivated land			Non-cultivated Inclusion			Grassland			Forest			Shrub			Others		
	1978/ 79	1985/ 86	An- nual change %	1978/ 79	1985/ 86	An- nual change %	1978/ 79	1985/ 86	An- nual change %	1978/ 79	1985/ 86	An- nual change %	1978/ 79	1985/ 86	An- nual change %	1978/ 79	1985/ 86	An- nual change %
High Himal	0.23	0.24	0.37	0.06	0.03	-6.77	26.40	26.42	0.01	4.63	4.63	-0.02	1.99	2.00	0.09	66.69	66.69	0.00
High Mountains	8.26	8.24	-0.02	4.97	5.00	0.08	17.23	17.16	-0.05	55.14	55.37	0.07	6.13	5.95	-0.42	8.27	8.28	0.02
Middle Mountains	27.51	27.53	0.01	14.97	15.02	0.03	6.58	6.26	-0.71	40.37	40.77	0.13	9.21	9.10	-0.18	1.36	1.33	-0.38
Siwaliks	13.73	14.26	0.56	2.93	3.13	0.96	1.10	0.85	-3.24	76.64	76.25	-0.07	1.66	1.54	-1.05	3.94	3.98	0.13
Terai	58.51	61.99	0.85	5.55	5.83	0.72	2.36	2.75	2.39	28.02	22.51	-2.81	0.07	1.42	291.84	5.50	5.50	-0.01
Total	20.12	20.69	0.40	6.69	6.77	0.16	11.91	11.83	-0.10	38.08	37.42	-0.25	4.68	4.79	0.33	18.51	18.50	0.00

Source: Neild, 1985

1971-72, 4.6% between 1971-72 and 1981-82 and 0.16% between 1981-82 and 1991-92. This rate of change in arable land between 1981-82 and 1991-92 was 4.0% in Mountains, 0.17% in the Hills and -0.19% in Terai. It has also been reported that arable land increased tremendously in Western Terai but it was decreased in Eastern Terai between 1981-82 and 1991-92.

Though the total area of all holding has been increasing but the average size of holding decreased from 1.11 ha in 1961-62 to 0.92 ha in 1991-92. It is reported that about 43% of holdings were less than 0.5 ha in size in 1991-92. The distribution of holdings is highly skewed. About 50% of holding operated only 16% of land (CBS, 1994).

Though the area as well as cropping intensity have been increasing (145% in 1981-82 to 175% in 1991-92). Increase in cropping intensity without providing enough plant nutrient replacements has been leading to a stagnant or declined yield. It is reported that inorganic fertilizer represents only 4% of the total plant nutrient replacements (Hrabovsky and Miyan, 1987). Productivity on the average for all grains was higher in the 1970s than in the 1980s (Banskota, 1992). The gap between the demand and supply of food has been increasing due to rapid increase in population (2.1% per annum) than the increase in agricultural production (1.8%). In the 1960s and 70s, Nepal used to export rice, but in recent years it has to import food grains in order to meet domestic demand. The production level meet only about 80-85% of the per capita food requirement in the country (World Bank, 1989).

Food situation in the country by physiographic regions is presented in Table 3. It reveals that the gap between the supply and demand of food has been increasing. The shortage of food increases as one goes from south to the north. Out of a total of 75 districts in the country, only 26 districts have a surplus in food grains.

Migration of people from the hills and mountains to Terai helped to some extent to reduce the increasing pressure in agricultural land due to shortage of food in the hills and mountains in the past. Migration from the Hills to the Terai remained very

Table 2
Landuse Change Reported from Different Places in Different Periods

Place	Ecological belt	Source	Period	Agri-culture	Forest	Shrub	Grazing	Others	Forest & shrub
Dhankuta (20000 ha)	Middle Mountain	Virgo & Subba, 1994	1978-1990	-0.05	0.53	0.89	-1.57	3.13	
Sankhuwashabha (13562 ha)	Middle Mountain	Bhadra, et al, 1991	1978-1990		-0.7				
Mahulikhola	Terai	Khanal, 1993		0.02	-0.06	-0.08			
Eastern Development Region	All	HMG/FINNIDA, 1996	1978/79-1985/86	-0.39	0.64				-0.25
			1985/86-1992		-4	12.86			-1.02
			1978/79-1992		-2	6.5			-0.59
Eastern Development Region	Hilly Areas	HMG/FINNIDA, 1996	1952-1978/79	-0.48	0.19				-0.35
			1978/79-1992	0.2	-2.16	4.42			-0.67
			1952-1992		-0.94	1.64			-0.43
Bamti-Bhandara (483 ha)	High Mountain	Ries, 1993	1965-1991	-0.93	6.05	3.14	-1.48		
Jhikhu khola	Middle Mountain	Shrestha & Brown, 1995	1947-1981	0.56	-1.63	5.86			
			1972-1990	0.58	2.99	-2.75	-3.37	-1.53	
			1947-1990	0.25	-0.71	2.54			
Nakkhu khola (41219 ha)	Middle Mountain	Tiwari, 1990	1954-1978	0.93	-0.2	-0.81			
			1978-1986	0.3	-5.52	11.66			
			1954-1986	0.79	-1.46	1.75			

Contd...

Contd...

Chakrimakri	Siwalik	Gurung & Khanal, 1987	1954-1979	69.85	-1.34		-0.15
Central Development Region	All	HMG/FINNIDA, 1995	1978/79-1985/86		-0.46	1.39	-2.08
			1985/86-1992		-2.35	-0.93	-1.03
			1978/79-1992		-1.3	0.28	-0.57
Central Development Region	Hilly Areas	HMG/FINNIDA, 1995	1956-1978/79	0.48	-0.35	-1.09	-0.67
			1978/79-1992		-1.58	1.92	-0.58
			1952-1992		-0.75	-0.17	-0.58
Upper Pokhara Valley (20339 ha)	Middle Mountain	Thapa & Weber, 1990	1957-1978	0.89	-0.56	-0.58	0
			1978-1988	0.23	-0.02	-0.69	-0.39
			1957-1988	0.69	-0.38	-0.59	
Karnali zone	Mountain	Bishop, 1990	1950-1972		-2.27		

high (3.5% in 1951, 6% in 1961, 9% in 1971, and 6.1% in 1981). Large areas of Terai forests were cleared for agricultural use. But now the Terai is fully saturated because of high volume of internal and international migration. The rate of extension of agricultural land into forest, shrub and grassland is likely to increase in the coming years even in mountain areas unless the present high rate of population growth is controlled, productivity of foodgrains is increased and an easy and effective market mechanism is developed.

Table 3
Food Situation in Nepal

Physiographic Regions	Total No. of Districts	No. of Districts with Food Surplus (1991)	Percentage of Supply to the Demand of Food	
			1985	1991
Mountains	16	4	88	66
Hills	39	13	83	79
Terai	20	9	122	99
Nepal	75	26	101	81

Source: Hrabovsky and Miyan, 1987 for 1985 and Manandhar, et al., 1994 for 1991.

Nearly 20% of the total land is under agricultural use. Out of which valley cultivation comprises about 58% of the total cultivated land and the remaining is hillslope cultivation. Sloping terraces from where the annual loss of soil and nutrients is very high comprise about 17% of the total cultivated land (LRMP, 1986). The land with slope less than $5°$ which is suitable for annual crops comprises only 17% of the total area in the country. Moderately sloping land $3-30°$ comprises about 15% of the total land (LRMP, 1986a) where terracing is mandatory for agricultural use and needs comparatively large amount of resources both labour and materials for the construction and maintenance of terraces. The remaining 68% of the total area in the country is either too steep or too cold or active flood plain which are not suitable for agricultural use (LRMP, 1986b). It shows that the extension of agricultural land in the future does not seem economically and environmentally viable.

There are some prospects to increase the level of production by increasing cropping intensity as well as productivity through the development of irrigation. Nearly 67% of the total agricultural land is irrigable (DI/HMG, 1990). Only 35% of the total agricultural land is presently irrigated. Only 13% of the total agricultural land is irrigated year round. In other words, the production of food grains in more than 87% of the total agricultural land depends on highly uncertain monsoon precipitation. So, efforts should be made to provide irrigation facilities in order to increase present level of production.

Forest and Shrubland

The process and the rate of change of forest cover in the country vary with time and space. (Neild, 1985) estimated that the forest land decreased by 0.25% per annum between 1978/79 and 1985/86 (Table 1). Very high rate of decrease in forest area (2.81%) was reported from Terai region whereas there was slight increase in the Middle and High Mountain regions. Forest land including shrub decreased by more than 30,000 ha per annum (-2%) in the Terai between 1964-65 and 1974-75 (Bajracharya, 1983a and Browning, 1974). Bajracharya (1983, 1983a, 1983b) reported a very high rate of deforestation (67,000 ha per year) in the Hills during the same period.

The major cause of such a high rate of deforestation in the country was for food and fodder. Studies on landuse/land cover change show a decrease in forest land in Terai by more than 30,000 ha per annum between 1964-65 and 1974-75, 13,000 ha between 1978-79 and 1985-86 and 8,000 ha between 1978-79 and 1990-91 (Table 4).

Recent studies in the Hilly Areas of Eastern and Central Development Regions showed an annual decline in forest areas including shrubland by -0.4% and -0.7% respectively between 1978-79 and 1992 (Table 4). Natural forest (excluding shrub-land) declined annually by -2.16% in Eastern Hills and -1.58% Central

Hills during the same period. Quite a few micro level studies such as in Bamti-Bhandara in Ramechhap district (High Mountain), Jhikhu Khola in Kavre district (Middle Mountain) and Dhankuta (Middle Mountain) show an increase in forest land in recent years (Table 4). This increase in forest land in Kavre and Dhankuta districts was due the increase in the number of trees in private field, efforts of development projects and the change in the structure of rural economy as the result of improved accessibility, market and off farm employment opportunities (Shrestha and Brown, 1995, Virgo and Subba, 1994, Gilmour and Fisher, 1991 and Mahat, 1985). It is also reported that though the forest area in watershed level in Jhikhu khola has been increasing after 1972 after the implementation of Nepal-Australian Forestry Project, the standing biomass in accessible forest plots has diminished from 614 trees to 386 trees resulting in a loss of 37% between 1989 and 1994. Household survey in the same area shows that about 71% people reported that it was easier to obtain fuelwood five years ago (Shrestha and Brown, 1995). After the construction of roads, the pressure on forest area has increased due to the introduction of new commercial activities such as brick factories and catering services. Another study in the same area indicate that increases in the number of trees in private land is not substantially helping to reduce the pressure on forest land due to highly skewed distribution of private land, negative association with the size of livestock and landholding and mass poverty in the area (Schmidt, 1992 and Banskota, 1992). A recent living standard survey also shows that more than 66% households rely on public forest, 12% on community managed forest and 19% on own land (CBS, 1996a).

Study in Khumbu region (High Himal) shows that the forest area was altered significantly in the past but not in recent years. However, standing biomass of Juniper forest was reduced by 36% in 22 years (Byers, 1987). Study in Karnali zone (High Mountain and High Himal) shows a decrease in forest land by 50% between 1950 and 1972 (Bishop, 1990).

Study of a 5 Village Development Committee in Shankuwashabha district indicates a decline in forest area by 0.7%

Table 4

Annual Change in Forest Area (ha)

	Nepal		Terai		Siwalik		Hills		Source
	Area	%	Area	%	Area	%	Area	%	
a. Forest area including Shrubland									
1964/65-1974/1975	-98,700	-1.7	-31,500				-67,200		Bajracharya, 1983
1964-1975			-29,667	-2					Browning, 1974
1964/65-1978/1979	-27,286	-0.4	-13,650	-1.7	-18,807	-1.1	5,171	0.1	Neild, 1985
1978/79-1985/86	-11,814	-0.2	-12,528	-2.1	-1,286	-0.1	2,000	0.1	Neild, 1985
1978/79-1985/86	-12,129	-0.2							
1978/79-1990/91			8,283			-1.3			HMG/FINNIDA, 1994
b. Natural Forest excluding Shrubland									
1978/79-1985/86	-14,114	-0.3	-16,614	-2.8	-957	-0.1	3457	0.1	Neild, 1985
1978/79-1985/86	-22,286	-0.4	-20,286	-3.5	-1571	-0.1	-429	-0.1	HMG/ADB/FINNIDA, 1988
1978/79-1985/86	-31329		-20,629		-1,628		-9071		Bhatta, 1992

per annum between 1978 and 1990. Decline in crown density class was also observed in about 15% of the total forest area during the same period (Bhadra, et al., 1991). Study in Nakhu watersheds (Middle Mountains) shows a very high rate of deforestation between 1978-1986 than in 1954-1978 due to increase in the pressure on forests after the construction of road and occurrence of natural disaster in the watershed (Tiwari, 1990). It has also reported a decline in crown density through the cutting of branches for fuelwood and fodder rather than felling of trees.

Very high rate of deforestation in the past (1957-1978) than in recent years (1978-1988) was reported from Upper Pokhara Valley (Table 4). The increase in the rate of deforestation in this area in recent years was mainly due to the fact that forests in the accessible proximity to the village and suitable for agricultural use had already been cleared before 1978. In addition to this increasing public awareness of environmental and economic consequences also resulted in the decrease of the annual rate of deforestation (Thapa and Weber, 1990).

The rate of deforestation is very high in Terai. It is mainly due to very high rate of population growth due to both internal and international migration, mass poverty, socio-economic inequalities, lack of permanent off farm employment opportunities, introduction of new commercial activities and other political and bureaucratic reasons (Ghimire, 1992). Study in Mahuli khola watershed (Siwalik and Terai) indicates a very low rate of deforestation as compared to the rate for the Terai as a whole but it shows an increasing pressure on forest land. Nearly 800 persons enter into forest land (2,000 ha) daily to collect fuelwood, fodder and other forest products. The purpose of 70% people collecting forest resources was for sale and most of them were landless people (Khanal, 1993).

The increasing demand of food, fuelwood, fodder and timber as the result of very high growth in population and infrastructures in the country has been causing for the increasing pressure in forest land. The fuelwood and fodder situation in the country is given in Tables 5 and 6. The gap between the requirement and the sustainable supply of fuelwood in the country is very high as compared to food situation. The ratio of supply to the demand of fuelwood for 1991 was only 39% which indicates a very high pressure on forest land for fuelwood. This pressure is very high in the Terai and the Hills than in mountains. Forestry Master Plan has estimated a deficit of 2.9 million tons for 1990/91 with a supply-demand ratio of 77% in the country (HMG/ADB/ FINNIDA, 1988). Hrabovsky and Miyan (1987) have shown the worst situation of fuelwood supply (20%) in the country. In order to fulfill the increasing requirement of fuelwood, the use of cowdung and crop residue has been increasing. More than 25% of the total surveyed household use dung, litter and thatch for cooking purposes (CBS, 1996a). As a result, crop and livestock sectors are highly affected. The sustainable supply of fodder is only 91% of the total requirement. Very high shortage of fodder is reported from the Hills. Though the supply of fodder is higher than the requirement in the Terai, but major portion of crop residue is used for cooking and heating. So, animal grazing in forest and shrubland is commonly practiced. Collection of fuelwood, fodder, litter and overgrazing have been leading to the rapid degrada .Ugradatiest land in the country. It has been estimated that natural forests have been degrading by 0.4% annually (Table 4). Thick forest with crown densities more than 70% comprises only 15% of the total forest area. Forest with crown densities between 40-70% comprises about 59% and the remaining 26% forest has crown densities between 10-40% (LRMP, 1986).

Table 5
Fuelwood Situation in Nepal

Physiographic Regions	Total No. of Districts	No. of Districts with Fodder Surplus (1991)	Percentage of Supply to Demand of Fodder	
			1985	1991
Mountains	16	2	34	72
Hills	39	0	14	35
Terai	20	5	24	38
Nepal	75	7	20(81)	39(77)

Note: Figures in parenthesis indicate the estimate made by Forestry Master Plan.
Source: Hrabovsky and Miyan, 1987 for 1985 and Manandhar, et al., 1994 for 1991.

As the result of rapid degradation of natural forest, shrubland has been increasing in the country. Large areas of shrubland in the Hills and Mountains have been converted into grazing and agricultural land.

Table 6
Fodder Situation in Nepal

Physiographic Regions	Total No. of Districts	No. of Districts with Fodder Surplus (1991)	Percentage of Supply to Demand of Fodder	
			1985	*1991*
Mountains	16	7	104	89
Hills	39	10	78	83
Terai	20	11	111	107
Nepal	75	28	92	91

Source: Hrabovsky and Miyan, 1987 for 1985 and Manandhar, et al., 1994 for 1991

Grassland

Grassland alone covers about 12% of the total area of the country. Grassland covers more than 26% of the total land in the High Himal and 17% in the High Mountain region. Percentage of grassland in other physiographic regions is less than 7% (Table 1). Most part of the grassland in the Siwalik and Middle Mountain have been converted into agricultural land. Large areas of highland pastures are inaccessible and the ecosystem so sensitive that both over grazing and under grazing lead to the reduction in the availability of forage which adversely affects the livestock population and its productivity. In a situation of low density stocking, shrub invades open enclaves reducing the availability of forage whereas in high density stocking, pastures are being converted into bare ground leading to high runoff and soil loss and eventually to terraces (Archer, 1990 and Watanabe, 1994).

The pressure on northern highland pastures particularly in Humla, Mugu, Dolpa, Mustang, Rasuwa, Dolakha and Shankuwashabha districts has been increasing after 1959 when the traditional open range grazing during winter in Tibet was closed for Nepali herds. It has put pressure not only on open pasture land but also on forest land (Miller, 1993, Archer, 1990 and Basnyat, 1989). Lack of knowledge of range ecosystem and pastoral system and poor extension and technology delivery systems are some of the problems associated with the proper management of highland pasture in the country (Miller, 1993).

Feed deficit in the country specially in the Hills has been leading to the increase in the size of overgrazed areas. As the result, annual transhumance movements in the west descend lower in the winters than they used to and more buffaloes from the Hills now go upto summer grazing pastures. The time required for gathering grass and fodder has also been increasing in recent years (HMG/ADB, 1993).

Above discussion on the process of landuse/ land cover change shows that many socio-economic and political factors are involved for the change in landuse and it seems that it is leading to unsustainability. It is in this context that the country should develop an integrated landuse policy for the overall economic and environmental development of the country.

References

Archer, A. C. (1990) Pasture and fodder development in the high altitude zone, Nepal, UNDP/FAO, NEP/85/007.

Bajracharya, D. (1983) Implications of fuel and food needs for deforestation: An energy study in a Hill Village Panchayat of Eastern Nepal, PhD Thesis, University of Sussex.

Bajracharya, D. (1983a) Fuel, food or forest? Dilemmas in a Nepali village. Working paper WP-83-1, the East-West Resource Systems Institute, East-West Center, Honolulu, Hawaii.

Bajracharya, D.(1983b) Deforestation in the food/fuel context: Historical and political perspectives from Nepal. *Mountain Research and Development*, 3(3), 227-240.

Banskota, K. (1992) Agriculture and sustainable development. In: *Nepal Economic Policies for Sustainable Development*, pp 51-71, Asian

Development Bank, Manila and International Centre for Integrated Mountain Development, Kathmandu.

Basnyat, N.B. (1989) Pasture and fodder development in high altitude zone, report on pasture and range land resources in Upper Mustang, UNDP, NEP/85/007.

Bhadra, B. Sharma, S. Khanal, N., Joshi, A. and Sharma, B. (1991) Environmental management and sustainable development in the Arun basin: management of natural resources, Volume 3. King Mahendra Trust for Nature Conservation, Kathmandu.

Bhatta, B.R. (1992) Management of forest resources. In: *Nepal Economic Policies for Sustainable Development*, pp 72-100, Asian Development Bank, Manila and International Centre for Integrated Mountain Development, Kathmandu.

Bishop, B. (1990) Karnali under stress: Livelihood strategies and seasonal rhythms in changing Nepal Himalaya, Geography Research Paper No. 228-229, Chicago University Press, Chicago.

Browning, A. J. (1974) Forest development, Nepal: marketing. Prepared for the government of Nepal. FO:DP/NEP/69/513 Technical Report No. 1. FAO, Rome.

Byers, A.C. (1987) A geoecological study of landscape change and man-accelerated soil loss: The case of the Sagarmatha (Mt. Everest) National Park, PhD Thesis, University of Colorado, Boulder, USA.

CBS (Central Bureau of Statistics) (1994) National sample census of agriculture, Nepal, 1991/92, analysis of results. NPC/HMG, Kathmandu.

CBS (Central Bureau of Statistics) (1996). *Statistical Pocket Book*, Nepal, NPC/HMG Nepal. Kathmandu.

CBS (Central Bureau of Statistics) (1996a). Nepal living standards survey, 1995-96, Statistical report, main findings, volume one. NPC/HMG Nepal, Kathmandu.

DI (Department of Irrigation)/HMG (1990) Master plan for irrigation development in Nepal. Canadian International Water and Energy Consultant in association with East Consult, Kathmandu.

Ghimire, K. (1992) *Forest or Farm?: The Politics of Poverty and Land Hunger in Nepal*. Oxford University Press, New Delhi.

Gilmour, D.A, and Fisher, R. J. (1991). *Villagers, Forests and Foresters*. Sahayogi Press, Kathmandu.

Gorkhali C.P. (1991) Biological diversity In: *Background papers to the National Conservation Strategy for Nepal*, Volume III PP 445-447. NPC/IUCN, Kathmandu.

Gurung, H. and Khanal, N.R. (1987) Landscape processes in the Chure Range, Central Nepal. Nepal National Committee for Man and the Biosphere, Kathmandu.

HMG/ADB (1993) Livestock Master Plan, Volume I: A strategy for livestock development, Kathmandu.

HMG/ADB/FINNIDA (1988) Master Plan for Forestry Sector, Nepal, Kathmandu.

HMG/FINNIDA (1994) Deforestation in the Terai districts 1978/79-1990/91. No. 60, Forest Research Survey Centre, Forest Resource Information System Project, Kathmandu.

HMG/FINNIDA (1995) Woody vegetation cover of the Eastern Development Region, 1992, No. 63, Forest Survey Division, Forest Research and Survey Centre, Kathmandu.

HMG/FINNIDA (1996) Woody vegetation cover of the Eastern Development Region, 1992, No. 67, Forest Survey Division, Forest Research and Survey Centre, Kathmandu.

Hrabovszky, J.P and Miyan, K. (1987) Population growth and landuse in Nepal "The Great Turnabout". *Mountain Research and Development,* 7(3), 264-270.

Ives, J.D. and Messerli, B. (1989) *The Himalayan Dilemma: Reconciling Development and Conservation,* The United Nations University, Routledge, London.

Khanal, N.R. (1993) Watershed condition, processes of change and its implication and management plans for the Churia and Terai Region of Nepal: A case study of Mahuli khola watershed in Saptari district, Nepal-German Churia Forest Development Project (ChFDP), Kathmandu.

LRMP (Land Resource Mapping Project) (1986) Land utilization report. Kenting Earth Sciences, Kathmandu.

LRMP (Land Resource Mapping Project)(1986a) Land system report. Kenting Earth Sciences, Kathmandu.

LRMP (Land Resource Mapping Project)(1986b) Land capability report. Kenting Earth Sciences, Kathmandu.

Mahat, T.B.S. (1985) Human impact on forests in the middle hills of Nepal. PhD Thesis, Australian National University, Australia.

Manandhar, M. Subedi, B.P, Malla, U.M., Shrestha, C.B., Ranjitkar, N.G., Pradhan, P. K. and Khanal, N.R. (1994) Study on population and environment in Nepal. Central Department of Geography, Tribhuvan University, Kathmandu.

Miller, D.J. (1993) Range lands in Northern Nepal: Balancing livestock development and environmental conservation. USAID/Nepal, Kathmandu.

Neild, R.S. (1985) Fuelwood and fodder: Problems and policy. Working paper for the Water and Energy Commission Secretariat (WECS), HMG/Nepal, Kathmandu.

Ries, J.B. (1993) Soil erosion in the High Mountain Region, Eastern Central Himalaya: A case study in Bamti/Bhandara/Surma area, Nepal. PhD Dissertation, Geowissenschftlichen Fakultat, Albert-Ludwings-Universitat, Freiburg I.Br.

Schmidt, M.G. (1992) Forest landuse dynamics and soil fertility in mountain watershed in Nepal: A GIS evaluation. PhD Thesis, Department of Soil Science, University of British Columbia, Canada.

Shrestha, B. and Brown, S. (1995) Landuse dynamics and intensification. In *Challenges in Mountain Resource Management in Nepal, Processes, Trends and Dynamics in Middle Mountain Watersheds* (eds. By H. Schreier, P.B. Shaha and S. Brown)(Proc. Workshop, April 10-12, 1995) International Centre for Integrated Mountain Development (ICIMOD), Kathmandu and International Development Centre (IDRC), Ottawa.

Thapa G.B. and Weber K.E. (1990) Managing mountain watersheds: the Upper Pokhara Valley, Nepal, Studies in Regional Environmental Planning HSD Monograph 22, Division of Human Settlements Development, Asian Institute of Technology, Bangkok.

Tiwari, D.N. (1990) Watershed modeling: Estimation of surface runoff and soil erosion rates. Msc Thesis, Asian Institute of Technology, Bangkok, Thailand.

Virgo, K.J. and Subba, K.J.(1994) Land-use change between 1978 and 1990 in Dhankuta district, Koshi Hills, Eastern Nepal. *Mountain Research and Development.* 14(2), 159-170.

Watanabe, T. (1994) Soil erosion on Yak-grazing steps in the Langtang Himal, Nepal. *Mountain Research and Development,* 14(2), 171-179.

World Bank (1989) *Nepal: Policies of Improving Growth and Alleviating Poverty.* Washington DC.

7

Spatial Modelling Approach for Water Resource Augmentation in Geo-environmental Perspective

D. Das

One of the greatest geo-environmental problems in the Tropics is widespread soil erosion. This is hampering agricultural production, vegetation growth and causing desertification. This type of land deterioration is the result of a complex sequence of human activities impacting on a fragile physical environment. Population explosion and associated increase in livestock numbers lead to forest clearance for fuel wood and an increase in use of dung for fuel, thereby reducing the vital inputs of natural organic fertilizer to the agricultural land. Declining agricultural production exacerbates the pressures on the remaining land, until overgrazing and loss of soil fertility eventually lead to famine and desertification (Maizels, 1990). Forest clearance also leads to exposure of the soils to erosion especially in areas of steep slopes and intense seasonal rains. Soil erosion of this scale leads to rapid silting of surface water. A thin layer of soil and steep slopes also promote rapid surface run-off, which in turn increases the risk of floods and reduces dry season base flow and recharge of groundwater reserves.

In India 75% or more of its population are relying on agriculture for their livelihood. Scientific and optimal utilisation of land and water resources is a pre-condition for sustainable socio-economic development in sound environment. In order to ensure sustainable development, it is very essential to prevent precious land resources from further deterioration through better landuse practices and to augment water resources both surface and subsurface against the vagaries of rainfall. In drought prone areas special emphasise has to be given to store and conserve rain water. The natural recharge studies have indicated that only 5 to 10% of the rainfall volume able is to percolate deep enough to augment the groundwater in hard rock areas, critically it depicts the emerging critical situations like drying up of shallow dug wells, poor yields from shallow boreholes (Muralidharan, 1997).

In the wake of population explosion, the deteriorating surface water bodies and constraints of harnessing safe groundwater, it is high time that all possible measures are taken to augment water resources. In the follow up studies it should be remembered that surface and groundwater system are interlinked. Due to convenience, low gestation period and smaller scale of investment, the demand for groundwater far exceeds the demand for surface water. This has been more geared up due to the availability of institutional finance for agricultural water supply. Lowering of water level in pre-monsoon as well as in the post-monsoon periods leads slowly to a large scale change in the vegetation pattern causing ecological imbalance in the area. Drying up of domestic tube wells during summer months also causes health hazards among the rural people.

The main objective of this article is to discuss how the water table condition can be improved mainly in the hard rock region and surface water storage increased through artificial recharge and water harvesting methods developed. Targeting for promising areas of groundwater development has also been discussed. In the above context a few case studies from drought prone Puruliya and Bankura districts of West Bengal have been elaborated watershedwise. Here the selected river basins are Kangsabati of Puruliya district and Sali of Bankura district where country rock is granite gneissis and average annual rainfall is 1120 mm and 1200 mm respectively.

Water Management Scenario

Uneven distribution of water resources is found almost everywhere in our country. This unevenness is spatial as well as temporal. For this reason, it is very common to have flood as well as drought in different parts of our country at any particular time. Drought is defined as a deficiency of precipitation over an extended period of time resulting in water shortage and extensive damages to crops resulting in loss of yield. It begins with the agriculture section (heavy dependence on soil moisture/water) as well as people dependent on source of water (Nagaranjan, *et al.*, 2000). In the olden days when there was less of population pressure and less interference with nature, the fresh water availability was enough. Adequacy of fresh water in river outside the saline belts and in tanks and wells were taken for granted. In the semi-desert terrains, there have also been attempts from very ancient time to draw fresh water by driving seepage galleries into groundwater bearing formations (Dasgupta, 1996). During 1967-1991, droughts have affected 50% of the 2.8 billion people who suffered from natural disasters and killed 35% of the 3.5 million people who lost their lives due to natural disasters (Rao, 2000). About 75 to 80% of the annual rainfall in major portions of our country occurs during the south-west monsoon in the month of June to September. While the Himalayan and Sub-Himalayan region receives water due to snow melting in summer, some regions in the north east and the east coast receive rainfall due to north east monsoon. Nearly one third of our country's area is said to be chronically drought prone. It has been observed that these areas generally have an annual rainfall of less than 750 mm. In these areas scarcity of water is observed very frequently. It calls for effective water management mainly by adopting evaporation control and rainwater harvesting (Desai, 2000).

In India, more than 90% of rural and nearly 30% of urban population depends on groundwater for meetings its drinking and domestic requirements. Groundwater also accounts for nearly 60% of the total irrigation potential of the country (Reddy, 1999). However, the demand for water is 3-4 times higher than the availability of both surface and groundwater resources. Hence, the gap between the resource available and its requirement is widening with increasing developmental activities. In recent years, it has been observed that a sharp decline in water table conditions has taken place due to over withdrawal of groundwater. Due to such mismanagement in groundwater exploitation, a scientific budgeting of groundwater is needed-based on its spatio-temporal distribution, its allocation for meeting competing demands like irrigation, industrial and domestic requirements.

In India the lower and upper bound values of the annual rainfall range are exactly the same as that of China. With the minimum rainfall occurring in the cold desert of Ladhak in Jammu and Kashmir as well as in the hot desert 'Thar' of Rajasthan while the maximum rainfall occurring in the Western Ghats as well as in Meghalaya (Desai, op.cit.). In India, in the recent years there are many successful stories related to water management like the artificial glaciers or 'Zings' of Ladhak in Jammu and Kashmir, the narrow wells or 'beris' of Jaisalmer in Rajasthan. All these successful stories not only testify our capability to achieve water conservation, but also give us a hint about the grim picture that we might face in the coming years if such attempts are not undertaken on a massive scale (Desai, op.cit).

In China, the annual rainfall varies all the way from less than 250 mm in the north west to about 2000 mm in the south and south west. This region has more than 80% of national water resources for a national area of little more than 35% supporting nearly 55% of the nation's population. The strong point of China is its retention of the traditional wisdom as well as adoption to modern techniques. In Israel, the average annual rainfall varies from about 150 mm in the southern desert to about 600 mm in the northern hilly region. Both these values are less as compared to both the extreme values of annual rainfall in the Indian state of Rajasthan. Yet, Israel has enacted laws and implemented programmes for better water conservation and management which ensure minimisation of wastage and thereby improving the water use efficiency. Some of the techniques used are rainwater harvesting and drip irrigation. As a result of such practices, Israel has been able to ensure water for drinking and other purposes for its citizens as well as to export citrus fruits (Desai, op.cit.).

Problem of Groundwater Development in Hardrock Region

Groundwater exploration in hard rocks tends to be very challenging because of their structural complexity and extreme variations at a very short distance. The major problems which are confronted while targeting for groundwater in the hard rock region may be listed as follows:

(i) Terrain usually shows undulating topography.

(ii) Highly auisotropic conditions of the aquifers.

(iii) Though the groundwater bearing zones could be generalised, there are established methodologies to pin-point well site location.

(iv) Large scale exploitation of groundwater and drilling of borewells at close proximity have rendered the groundwater go great depth. Causing failure of earlier groundwater structures.

(v) Movement of groundwater is an enigma.

(vi) Maps derived from satellite data gives information about joints and fractures,weathered zone etc. however, still it is difficult to gain detailed hydrogeological information of the terrain.

(vii) Fragmentation of cultivable lands into smaller units leading to closely spaced wells.

(viii) Absence of legislation to regulate the usage of groundwater has led to unhealthy competition towards its withdrawal.

Spatial Modelling Approach Towards Groundwater Development

A thorough local and regional hydrogeological understanding is necessary for an effective development of groundwater resources, especially in areas where bed rock has low primary porosity and where the intersections of secondary structural features is crucial for stressful exploration (Babu Rao, *et al.*, 1997).

During the last two decades, satellite based remote sensing has been proved to be efficient in mapping the suitable areas for groundwater prospecting in different scale. During the first developmental stage of hydrogeologic remote sensing application Landsat and IRS series of satellite were used mainly for carrying out regional level mapping on 1:1250,000 scale. Under National Drinking Water Technology Mission, the Department of Space, GOI, with the active support of different organisations has prepared district-wise hydrogeomorphological maps indicating prospective groundwater zones on 1:250,000 scale covering all the 446 districts of our country during 1987-92 period (Reddy, 1999). Again in the recent years under Rajiv Gandhi National Drinking Water Mission, NRSA (DOS, Govt. of India) has been entrusted to develop groundwater prospective zones in 1:50,000 scale. In India the IRS -- WiFS based detailed monitoring of drought was operationalised for Andhra Pradesh in 1998, and subsequently extended to Orissa and Karnataka. IRS WiFS data having 188 m. spatial resolution is useful for providing information about surface water spread, sowing etc.

By using the LISS-III and PAN merged data of IRS-IC/ID satellites, it has become possible to prepare maps upto 1:15,000 scale showing the surface water bodies, groundwater irrigated areas, canal commands etc. as small as 0.25 hectares (Reddy, op.cit.). Geomorphology, relief of the terrain, weathering status, fracture and joint pattern, lithology, soil type, and landuse can be deciphered from the satellite data with limited field checks. All these features are directly or indirectly influence groundwater occurrence, movement and accumulation. The information integrating lithology, land form, structure, land cover/landuse and aerial aspects of the drainage basin provide very useful clue for groundwater targeting and siting artificial recharge structures. By combining the remotely-sensed information in the form of various thematic layers in a GIS environment with adequate field data, particularly, well inventory and yield data, it is possible to arrive at prognostic models to predict the ranges of depth, the yield, the success rate and the types of the well suited to various terrain under different hydrogeological conditions. In order to assess the groundwater prospect by qualitative modelling, data on geomorphology, geological structures, and lineaments would also be required in addition to weathered zone thickness, saturated zone thickness and yield in wells. Lineaments being, line data has to be converted into spatial data by filling the lineament density on grid map.

Case Studies

In the present study an attempt has been made to appraise the landforms specially related to denudation and tectonics. Delineation of landforms is done on the basis of (a) landcover (b) water resource so that land and water resources of the area can be managed in a better way in the upper catchment area of Kangsabati river basin, and Sali river basin of Bankura and Puruliya districts respectively of West Bengal.

Kangsabati River Basin

To achieve such objectives, geomorphological study of the area has been carried out on the following aspects. For convenience of study, the area has been divided into two major physiographic divisions namely the (i) Denudational hills and Plateau tops, (ii) Low lying and slightly undulating plains. Various major and minor forms like residual hills, stream valley, palaeochannel on such divisions have been demarcated (Figure 1). Interrelationship of various forms within such physiographic divisions have also been studied from the form process relation and the stage of development. Nature and extent, distribution of planar surfaces have been specifically studied as they might provide important clues for the evolution of landscape (Figure 2). Geological mapping of the areas incorporating regional lithology and structure has also been carried out. Geomorphological or landform mapping has been performed with the help of remote sensing data (Landsat MSS-124 -- 1: 2,50,000, 1: 50,000), and from 1: 2,50,000-1: 50,000 scale SOI topographic map No. 73 I/3, I/4, I/7) and field checks. Composite map (Figure 2) using GIS technique has been developed. The drainage network on the 1:50,000 scale topographical maps have been reconstructed.

Figure 1
Composite Map of the Upper Catchment of Kangsabati River,
Purulia District, West Bengal

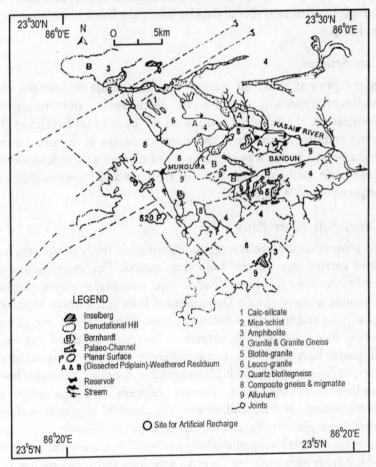

23°30'N
86°0'E

23°30'N
86°0'E

N

O   5km

B   3

4

6

1

KASAI RIVER

8

6   A   4

4

8

8   A

9

BANDUN

MURGUMA   B   9

9   8

8   4

4

8   5   6

520 P

4

8

7

8

9

3

9

LEGEND

Inselberg
Denudational Hill
Bornhardt
Palaeo-Channel
Planar Surface
(Dissected Pdiplain)-Weathered Residuum
Reservoir
Streem

1 Calc-sillcate
2 Mica-schist
3 Amphibolite
4 Granite & Granite Gneiss
5 Blotite-granite
6 Leuco-granite
7 Quartz blotitegneiss
8 Composite gneiss & migmatite
9 Alluvium
O Joints

A & B

O Site for Artificial Recharge

86°20'E
23°5'N

86°20'E
23°5'N

Observation and Results

Ajodhya plateau, the main upland of the study area is situated in
the southern portion. Another, relatively smaller raised landmass
than the earlier one, is present on the northern side of the study
area. Excepting those, various small residual hills are also present in
the study area. Geomorphic processes both exogenous and
endogenous, lithology and structure are responsible for sculpting
the present day landscape. There are three major planar surfaces

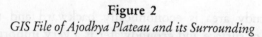

Figure 2
GIS File of Ajodhya Plateau and its Surrounding

identified at three different elevations in the study-area. They are situated at the elevations like 560 m, 340 m, and the third one (which has got maximum aerial extension in comparison to the others. The third planar surface is situated within two uplands (mentioned earlier) river Kangsabati flows through the almost central portion of the third planar surface (Figure 2). The width of the Kangsabati valley is small in comparison to the third planar surface. Several inselbergs have developed on the Ajodhya plateau and also in the lowermost planation surface (Figures 1, 2). Gradual opening up of the joints has developed the valleys and of course, the slope retreat phenomena has helped the formation of valleys. The presence of gorges, waterfalls and nick points suggest neo-tectonic activity. The role of multicyclic erosional processes in landsculpture development was suggested by the presence of paleodrainage system (Figure 1). The accumulation of boulders in the river bed indicates the recent erosional activity. Development of lateritic soil in the plateau area indicate the warm humid climate during their formation.

The study area is economically backward and the proper utilisation of surface and groundwater is a very important aspect. A map has been produced showing geology, landform with hydrogeomorphic features. The soils of the study area is of residual type which have been derived from the weathering of the precambrian granties, gneiss and schist. Lateritic soils are present in the upland areas and reddish clay loam or simply reddish clay are common in the valleys.

Groundwater resource investigation in the study area has revealed three hydrogeomorphic zones namely (1) denudational hills and plateau (2) palaeochannels and weathered residuum and (3) low-lying plains comprising precambrian crystalline (Figure 1). Ground water resource potential appears to be poor due to high amount of surface runoff and evapotranspirational loss. Such loss is mainly due to the (1) rocky nature of the terrain, (2) moderate to high gradient of the slope for most parts of the area, (3) uneven distribution of rainfall during the year.

Dutta and Banerjee (1979) tried to evaluate the groundwater resources of the area and observed its further scope of development. Remote sensing and GIS techniques aided by field checks give very fruitful information in identifying various special features like water bodies, drainage pattern, landform features etc. (Das 1990, Short and Lohman, 1973) which in turn give the information about subsurface water conditions. Thickness of the saturated zone within the weathered profile of planar surfaces varies from 3-4 m. as measured in some dug wells. Dug wells in such aquifers show poor yield (Dutta and Banerjee, op.cit.) groundwater yield from the terraces and palaeo channels appear to be low to moderate because of their limited spatial extent. Dutta and Banerjee (1979) observed considerable yield of water (600-5000 l.p.h.) in some bore wells drilled in the fractured granites in the area. Groundwater potentiality of the piedmont alluvial fans is comparatively high, which is evident from the presence of a good number of perennial tanks (Figure 2) and considerable water flow along the major stream courses. Groundwater potentiality in the low lying and slightly undulating region is not promising enough due to its surface characteristics (Mukherjee and Das, 1989). Several tanks have been built for storage of surface water as these may be the suppliers of drinking and irrigation water during dry period (Figure 2). Groundwater potential appears good where alluvial fan abuts against the flood plain of Kangsabati river. The alluvial deposits are good sites for groundwater potential and as well as suitable sites for agriculture. Maximum possible paleodrainage systems have been reconstructed through remote sensing and field checks (Figure 1).

A zone of upwarping (trending NW-SE) was detected when a comparative observation was given between present day drainage system and the paleodrainage system. Subsurface water resource investigation should be given special emphasis as the groundwater is confined both in hard rock and the alluvial region. Surface water resources should be developed mainly for irrigation and groundwater recharge purpose. Suggested sites are the confluence of 3rd and 4th order steams where alluvial sediments of variable thickness (2-15 m) underlain successively by weathered residuum and basement crystallines with intersecting fracture occur (Mukherjee and Das, 1996).

Sali River Basin, Bankura District, West Bengal

In the study area the basement gneiss and schist are much weathered at the surface and are generally mantled by a soil cover, the thickness of which vary from place to place. At places granite-gneiss and hornblende schist stand as moulds and ridge being resistant to weathering. The compact rocks are well foliated and have well defined joints. The four sets of joints are ENE-WSW, ESE-WNW, SE-NW, SSW-NNE. Foliation dip 40°-50° NE. The joints do not appear to show any regularity of spacing in a particular type of rock but the common spacing is between 1.5 to 2 m. Vertical joints are widely spaced and are not common.

Geomorphologically the area exhibits stepped sequence of three terraces of which the oldest is capped by laterite and/or yellowish brown sticky clay with kankar. Sand beds with a depth of 8 m. are correlatable with the oldest terrace. Repeated cycles of oxidised silt, incoherent gray to brown clay, coarse to fine grained brown sand occur below the capping of laterite and sticky clay with kankar. There is no gully erosion in laterite. The laterite in fact forms a protective cover, which is very resistant to both chemical as well as mechanical erosion.

Observation and Results

Drainage is by far the most potent in shaping the landform. Following the natural ground slope all the rivers maintain SE direction. The rivers are ephemeral and maintain a little amount of surface flow during dry season. The main channel of Sali contributes relatively more water in the basin.

. Third order subbasin have been studied in context of their large dimension which facilitate recognition of lithology structure and nature of sediment in such basins. Since the area comprises mainly of erosional landscape the development of basin is mainly due to the slope retreat process jointly with the action of structural and lithological control.

The development of tributary system appears independent of the major rivers and not guided by them in straight course, right angle bends in a system of parallel tributaries of the same order just opposite to the principal river suggest their entrenchment along some lineaments, a number of tributaries aligned parallel to each other suggest a zone of weakness, a minor fault or a fracture zone. The straight tributary course in few cases, are tectonic depressions occupied by rivers and changed into river valleys. The longitudinal profile of the valley has no slope vertically and occasionally the valley floor appears like a flood plain. It has to be clear from deposits, gentle slope and the graded profiles, that the river has no flood plain as situated in the Gangetic basin. The graded profile on the other hand are quite entrenched in the lateritic plateau.

The relative rate of valley incision and valley widening vary in different parts of the longitudinal profiles of the river. Very close to the source both valley widening and downcutting are active. The former as a result of recession of slope and the later because of steep gradient. The river profile is not effective to drain the catchment of the large river and during heavy monsoon rains often experiences floods. The silts derived from the uplands form the silt plains in this area. On the lateritic plain neither downcutting nor valley widening takes place at a rate noticeable enough. The general profile of the river does not permit any further incision and the lateritic cover is almost immune to chemical weathering and does not succumb to the process of weathering and erosion. It must, however, be cleared that much of the nonlateritic area is flat and cultivated.

It has been observed in areas having low inclination (30° to 45°) of joints, the well contain sufficient water throughout the year and show low seasonal fluctuation of water table whereas in areas having higher inclination of joints (more than 60°) show higher recession of water level with the approach of summer. The well that did not tap the entire thickness of the weathered zone was found to contain very little or no water during the summer months. The range of fluctuation was large in the well where frequency of occurrence of joints and the thickness of weathered zone encountered were less. The mode of occurrence of ground water varies from unconfined condition in hard rock and lateritic areas to semiconfined condition in recent alluvial plain. As regard to ground water potentiality, low potential areas are underlain by pre-Cambrian rocks. The aquifer was made up of weathered residua and the fractures/joints underlying it. Low to medium potential area occupies transitional zone between hard rock and young alluvial terrain. The aquifer was made up of thick weathered residuum and semiconsolidated sand underlying the lateritic hard crust. Moderate to high potential areas were underlain by multi aquifer systems made up of unconsolidated sand (younger alluvium).

GIS Analysis

Taking several pertinent thematic layers, viz. drainage, lineament, palaeochannel, thickness of buried pediment etc. in GIS environment (visual interpretation of IRS-1B LISS-II, scale 1: 50,000 data using ARC/INFO software package), a suitable

Figure 3
Sali River Basin of Bankura District
(Map Showing Sites of Artificial Recharge Structures)

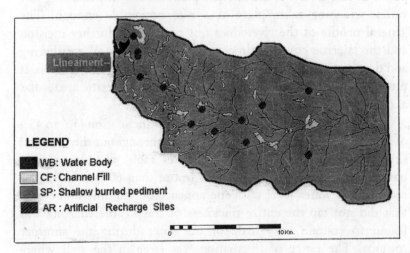

LEGEND
- WB: Water Body
- CF: Channel Fill
- SP: Shallow burried pediment
- AR : Artificial Recharge Sites

composite map (the study area is bounded by longitude 87°3' to 87°9'E and latitude 23°22' to 23°26'N) has been derived (Figure 3) showing sites for artificial recharge of groundwater.

Probable Recommendations

High
Highly favourable zones for recharge includes buried pediments (shallow) having moderate weathered thickness (20 m.), hydrologic soil group B with moderate infiltration characteristics, moderate runoff and moderate slope (2%). Suitable recharge structures are percolation ponds and check dams.

Moderate
Moderately favourable zones for recharge include valley fills with moderate thickness (16-20 m.), hydrological soil group B with moderate infiltration characteristics, moderate runoff and comparatively higher slope (2-5%). Suitable recharge structures are percolation ponds and check dams.

Less

Less favourable zone for recharge include buried pediment (deep) and flood plains having moderate weathered thickness (10-16 m.), hydrologic soil group C with less infiltration characteristics, moderately high runoff with high slopes (5-8%). Suitable recharge structures are recharge pits and recharge trenches.

Conclusion

Composite maps which have been generated using various thematic layers derived from visual interpretation of satellite data (IRS 1B and Landsat-MSS), toposheet interpretation aided by field checks have been analysed in GIS environment. Planar surfaces and other landform features are clearly demarcated through GIS techniques, which proves the effective role of spatial modelling towards the appraisal of landforms. Again crossing/analysing of various pertinent thematic layers reveal proper sites for construction of artificial recharge structures. Since the study area falls in a hard rock region and where the surface runoff is high, it is advisable to store the water (from natural precipitation) and adequate protective measure should be taken so that minimum evaporation loss can take place. Over exploitation of groundwater depletes water table condition. Surface storage of water shall be used in conjunction with subsurface water to manage water resource judiciously.

References

Babu Rao, P. and Seshu Babu, K. (1997): Application of remote sensing for groundwater targeting in part of Andhra Pradesh. *Proceedings of International Conference on Management of Drinking Water Resources,* CLRI, Chennai, India, pp.398-405.

Das, D. (1990): Satellite remote sensing in subsurface water targeting *Proceedings of ASCM-ASPRS Annual Convention* (Denver, U.S.A.), pp.99-103.

Dasgupta, A.B. (1996): Aquifer characteristics and groundwater management in eastern India. *Proceedings of the Seminar on Aquifer Characteristics and Groundwater Management in Eastern India* (Jadavpur University, Calcutta), pp. 1-2.

Desai, V.R. (2000): Towards better management through GIS/other strategies. *Pre-conference Proceedings CSDMS Map India 2000* (New Delhi), pp. 19-21 (WRM).

Dutta, D.K. and Banerjee, I. (1979): Groundwater development potential of Puruliya district, West Bengal, *Tech. Report CGWB, E Region*, Scr.F.

Maizels, J. (1996): Hydrogeological survey of water resources meeting the needs of aforestation projects in Ethiopia, *AGID News* No.63/64, pp. 13-16.

Muralidharan, D. (1997): Role of Water Quality in Artificial Recharge through Borehole Injection Technique, *Proceedings International Conference on Management of Drinking Water Resources* CLRI, Chennai, India, pp.139-150.

Mukherjee, A.L. and Das, D. (1989): A study on the development of basin and their hydrogeomorphic features in and around Ajodhya Plateau, Eastern India. *Proceedings of International Symposium on Intermontane Basins: Geology and Resources* (Chiang Mai, Thailand) *AGID*, pp.409-417.

Mukherjee, A.L. and Das, D. (1996): Morphometrical analysis for watershed management plan with special reference to aquifer recharge: A case study. *Proceedings of the Seminar on Aquifer Characteristics and Groundwater Management in Eastern India* (Jadavpur University, Calcutta), pp.143-147.

Nagarajan, R. and Mahapatra, S. (2000): Land-based information system for drought analysis. *Pre-conference Proceedings CSDMS Map India 2000* (New Delhi), pp. 11-13 (NDM).

Rao, D.P. (2000): Disaster management. *Pre-conference Proceedings CSDMS Map India 2000* (New Delhi), pp. 1-5 (NDM).

Reddy, P.R. (1999): Satellite data in groundwater resource budgeting and management for sustainable development; Pre-conference Proceedings, *Geoinformatics Beyond 2000* (IIRS, Dehradun), pp.278-283.

Short, M.N. and Lohman, D.P. (1973): Earth observation from space. Outlook for the Geological Sciences. Report. No. V-650-73, 316, Godard Space Flight Centre, Greenbelt, Marryland, p. 115.

8

Biogeography of Termites:
A Case Study of Mounds around Pune City

R.G. Patil

A number of geomorphologist constantly focused on explanation of evolution of landforms, involving both the explanation of landforms and also of earth surface processes. The influence of landforms on the distribution and development of plants, animals are studied by Biogeographers, while the influence of plants and animals on the earth surface process and the development of land forms is one of the neglected issue by the geographers, particularly geomorphologists. The biological aspect in the geomorphological process has been divided into passive and active effects. The active effects include the process that modify the soil as a growth medium by changing the status with respect of air, water, and nutrition and secondly, it involves the process of physical movement of soil particles, and largely contribute to profile homogenisation or horizonation. It includes the processes like producing special soil factors, regulation of soil erosion, forming and destroying soil fabric; mixing soil horizons; creating micro-relief by mounding and excavation, forming and filling soil voids etc. In the creation of micro-relief earthworms, termites, cicadas, scarabs have established a role with geographic factors and their features which have attracted the biogeographer and the biogeomorphologist.

Review of Literature

From the point of geomorphology, it is an accepted fact that Mounds are micro-relief features and all the features formed by vertebrates and invertebrates should be studied as a separate branch, i.e., biogeomorphology. Some of the scientist have questioned why geomorphology studies have begun to consider ecological factors in more depth, but since the earth sciences are continually developing and changing, there is an emergence of new problems to study. Krumbein (1978, 1983) and Ehrlich (1981) focused on the importance of the development of biogeomorphological studies which are responsible for the growth of biogeochemistry and geomicrobiology. Such studies of micro-relief are concern of micro-organisms and lower plants, which has been much neglected by biogeographers and ecologists. Viles H.A. (1988) remarked that biogeomorphology is not intended to be a long-lived term describing a separate field of research. But there are two linked foci of interests that can be identified in biogeomorphology, (1) the influence of landforms/geomorphology on the distribution and development of plants, animals and micro-organisms (2) the influence of plants, animals and micro-organisms on earth surface processes and the development of landforms, i.e., the influences are interdependent with respect of environmental equilibrium or change. Therefore such studies like role of termites and earth worms are to be studied by biogeographers, geomorphologists, pedologists and ecologists.

Termites are lower in the evolutionary scale, and nearly three thousand species are known, which are the member of Isoptera order and about four-fifth species belongs to the family Termitidate (Harris, 1961). Though termites are fierce, sinister and often repulsive (Maeterlinck, 1972) they are remarkable for having been adapted to living in highly organised communities for as long as 150-200 million years (Skaife, 1955). Termites are the most tenacious, most deeply rooted, most formidable of all the occupants and conquerors of globe. According to the UNESCO (1979) termites occur in great number from 2.3 million per hectare in Senegal to 9.1 million per hector in the Ivory coast.

Termites, generally known as 'white ants' are social insects and polymorphic. There are two basic types of habitats, soil and wood. The soil inhabiting termites build mounds above the soil surface while wood inhabiting are mostly arboreal, constructing their nests inside or outside of tree-trunks or branches. Termites are beneficial in their rapid turnover of organic matter in the ecosystem, which is done by the breakdown of cellulose and returning it to the soil.

The majority of termite species are found in tropics but the distribution is wider up to 45° north and south and the numbers decrease rapidly with increasing altitude and latitude. Nearly 3000 species are known from the world and more than 200 species found in Indian subcontinent, out of which 40 have been recorded as damaging crops or habitations which belongs to family *kalotermitides, hodotermitides, rhinotermitides, stylotermitides*. In India 38 species of genus *odontotermes* is dominant termite group.

The individuals of termites are recognisable into four caste forms which are: primary productive (king and queen), supplementary reproductive, soldiers and workers. The reproductive, consist of winged males and females and main function of both primary and supplementary reproductive is to keep the colony at optimum population by producing egg (male and female). The main function of soldiers is to protect the colony by physically biting or by secreting toxins and stop the entry of enemies by blocking entrance hole. The workers carry-out chores, work like foraging, feeding the young, tending the eggs, repairing and enlarging the nest and internal galleries.

Termite mounds are the most striking manifestation of termite activity and are enduring structures. Two types of mounds in Kenya have been identified are (Darlington, 1984) namely Bissel type and Marigat type. The main difference, lies in their ventilation system, which serve to cool a large, highly centralised nest. Both have underground foraging passage and outwash pediment. Bussel type has a continuous basal hump with more than two funnel openings, but Marigat type have generally only one funnel opening. The population was suspected to vary with seasonal conditions (Cowan, 1985).

Some of the studies carried out by zoologists, ecologists and geographers and edited by Viles H.A., like the tropical areas are studied by Darlington (1984), Cowan (1985), and in India, by Mohindra and Mukerji (1982) studied the mounds from the point of view of fungal ecology. Since the data of mound density, seasonal effects, variation in altitudinal zonation, factors affecting the mound, are not available on large scale; therefore the study of mound is considered from the biogeographical view point.

Objectives

The objectives of the present study are (a) to assess the role of geographical factors on the mound, and (b) study the characteristic which controls the mound formation and compare the soil characteristic between mound and surrounding area.

The Study Area

The study area is 0.33 sq. km. and located between Pune and Pimpri-Chinchwad Corporation at 18°30"45' N. latitude and 73°51"32' S. longitude, but the area is under Central Government and acquired by military. Before 1985 the area was kept fallow and converted by seasonal grass only in rainy season. Every year, this area was given for grazing purpose to the local people on auction basis by military authorities. After 1985, the Pimpri-Chinchwad Corporation planted the area and after 1989 the termites started construction of mounds.

Data Source

The data for the study is based on observation from last 20 years and primary data is collected from the field work on February 19, 2000, April 21, 2000 and August 15, 2000.

Methodology

Various micro-relief changes in the landforms were recorded and categories were identified, delineated and mapped. Special distribution of mounds were marked by chain and tape survey. For understanding the morphology of mounds the area was studied, three times, and 20 samples were collected with their north, south, east, west length and height. The soil samples of mounds and the surrounding area were collected for mechanical analysis.

Discussion

The study area is 0.33 sq. km., and situated at 616.66 m asl, having average rain fall 714.7 mm., average temperature of summer is 40° C., winter is 10° C., and of the 42 mounds which were found on February 19, 2000, 20 samples were collected for further discussion. The area was again surveyed on April 21, 2000 and August 15, 2000 for further study. On August 15th fine new notable mounds were observed, which had an average height of 20 cm. above the ground and 40 cm. vein below the ground level. Some 17 mounds show the signs of construction and other 14 holes of crab or other animals were noted. Therefore the construction is not continuous for a year but seasonal. Mainly the construction is from August to February, i.e., absent in rainy season and repairing of mounds carried out from March to May, which are hot months of the area and having temperature more than 30°c.

Before the plantation (1985), ecosystem of an area was under stress and low biomass, small structure, and low productivity; which have not supported the termite life and construction of mound. After the plantation (1989) biomass was increased, and construction of mounds started but was restricted to the morphology of area (Figure 1). The studied area is divided into two parts, one hilly and other plain. The hill is 10 meter high and having 30° slope, the plain area is divided into areas having stony outcrop, area under forest (covered with trees having height above 2 meters) and fallow land, where short grass is growing. Mounds found in the area are either in the forest or are along the gullies and are fully absent on or around stony areas, i.e., mounds tend to be rare on shallow soils and stony areas. They are not found even in the seasonal channel area but out of 20 samples, 8 are constructed near the flood channel range which shows soil drainage is important factor for construction.

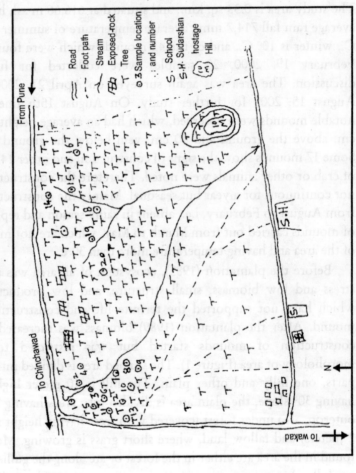

Figure 1

Location of Sample Site and Morphology of Area

The intensity of mound use varied considerably, three mounds were used for rubbing by animals and in one a frog was present, but most of mounds were covered by grass in rainy season.

Geographically, 75% of mounds are constructed in north-south direction, i.e., the length of north and south side is more than that of east and west (Table 1). Out of 20 samples, five are not in north south direction because three are constructed around Babul tree (Acacia) (sample no. 5, 6 and 20) and two are found to be abandoned (sample no. 15 and 17) having cracks 0.5 cm. to 2.5 cm. in width and 10 cm. to 35 cm. long.

Table 1

Length, Height and North-South Construction of Mounds

Sr. no.	North M.	South M.	East M.	West M.	Height M.	Direction and remark
1.	2.8	2.5	2.7	3.3	1.5	N-S
2.	2.9	2.8	2.4	2.5	1.2	N-S; Construction in progress
3.	1.4	1.2	1.3	1.2	0.7	N-S; Construction in progress
4.	1.6	1.6	1.3	1.4	0.7	N-S
5.	1.6	1.7	1.8	1.8	1.2	Acacia arabica to west
6.	2.3	2.1	2.3	2.4	0.9	Acacia catechu to west
7.	2.5	1.6	1.8	1.8	0.4	N-S
8.	2.0	1.8	1.9	2.0	0.7	N-S
9.	1.4	1.3	1.4	1.3	0.6	N-S
10.	2.7	2.6	1.7	1.6	0.6	N-S; Construction in Progress
11.	2.5	2.0	1.8	2.0	0.9	Zizyphus jujuba in middle
12.	2.7	2.2	2.1	1.9	0.8	N-S
13.	3.5	3.4	3.3	3.5	1.4	N-S
14.	2.2	2.1	2.0	2.3	1.2	N-S; ant eggs & dry abdomen
15.	3.9	3.7	4.1	4.0	1.5	Abandoned; 0.5_2.5cm. cracks
16.	2.6	2.3	2.5	1.3	0.8	N-S; Construction in progress
17.	3.0	2.7	3.4	2.9	1.5	Abandoned; prints of animal horn
18.	2.6	3.0	2.8	2.5	0.8	Acacia; twin mound; frog present
19.	1.5	1.9	1.8	1.4	0.7	N-S; Acacia to west
20.	2.2	2.4	3.3	2.3	0.8	Acacia catechu to west
Total	47.8	44.8	45.6	43.3	18.4	
Avg.	2.4	2.2	2.3	2.2	0.92	

Secondly the north side is more than that of south, except (sample no. 5, 18, 19 and 20), because the western side is attached to a tree. This indicates that three is some relationship with direction and construction; one may suggest that, for avoiding the heat and for maintaining the temperature of mound, termites may construct wider sides of north and south. The line graph (Figure 2) shows a constant relationship, which indicate that while constructing the mound, all direction length and height are balanced. The logarithmic graph (Figure 3) does not show any relationship, because it may have some relation with the vein below the ground and foraging passages which were not considered here. For this a detailed study and large amounts of data may be required. The mounds observed in the area are of Bissel type, i.e., more than one ventilation.

The mechanical analysis of mounds and surrounding soils clearly shows that 85% of mounds have more pH than the surrounding soil, three samples (no. 7, 8, 13) have less pH out of which two have 0.1 and one have 0.2 or less pH. The sample no. 13 is under shadow of trees naturally the evaporation rate is less. The average pH difference is +0.37 and out of 17 samples, 9 samples have more pH than average. The highest difference +1.2 is found for sample no. 14, which is the only mound where white eggs and dried parts of termite abdomen were observed, and laboratory analysis shows nearly 5000 dry abdomen were present in 10 gm. soil. It indicates that the mound attacked by termite after and before the sampling days. All the samples having more pH than average are located near the seasonal gullies out of which one to five are along the gully which has a source of limited drainage water, (four families drainage water) and are in an open area, because of this location the capillary action brings solution from a high ground water table and the mounds acts as a wick from which evaporation takes place. The other remaining samples which have more pH than surrounding soil but less than the average are away from the gullies and are located on gentle slope.

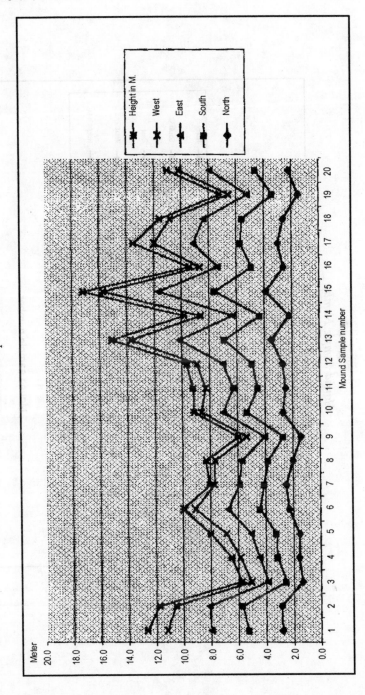

Figure 2
Mound Sample Number

Figure 3
Mound Sample Number

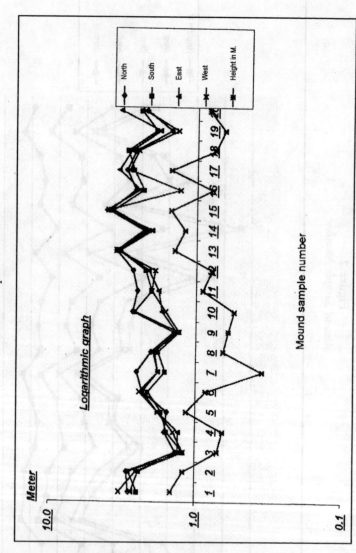

A clear manifestation of mechanical soil analysis is provided when one considers the sand, silt and clay contents of the mounds (Table 2). In all the samples (except 11 and 12) the clay content of mounds is elevated in comparison with surrounding soils. The mean clay content of mounds is 17.35%, while that of surrounding soil is 13.5%. The percentage of silt is more in all mounds than sand and clay, which refers to the ability of termites to mobilise particular soil particles and may be important in the development of prominent features. The upper part of silt layer is made up of very fine deposits.

Table 2

pH Values and Soil Composition of Termite Mounds Compared to Surrounding Soil

Sr. no.	pH values			% of Mound Soil			% of Surrounding Soil		
	Mound	Soil	Diff.	Sand	Silt	Clay	Sand	Silt	Clay
1.	7.3	6.9	0.4	11.0	76.0	13.0	73.0	16.0	11.0
2.	7.9	7.2	0.7	6.0	78.0	16.0	79.0	11.0	10.0
3.	7.4	6.9	0.5	5.0	70.0	25.0	67.0	17.0	16.0
4.	7.4	6.7	0.7	14.0	74.0	12.0	76.0	12.0	12.0
5.	7.1	6.6	0.5	4.0	81.0	14.0	59.0	31.0	10.0
6.	7.5	7.2	0.3	23.0	61.0	20.0	72.0	15.0	13.0
7.	7.2	7.3	-0.1	3.0	79.0	18.0	61.0	28.0	11.0
8.	7.1	7.3	-0.2	19.0	52.0	29.0	62.0	17.0	21.0
9.	7.3	7.2	0.1	36.0	42.0	22.0	51.0	30.0	19.0
10.	8.1	7.9	0.2	9.0	63.0	28.0	64.0	13.0	23.0
11.	7.4	7.1	0.3	10.0	84.0	6.0	73.0	17.0	10.0
12.	7.6	7.3	0.3	9.0	86.0	5.0	80.0	12.0	8.0
13.	7.7	7.8	-0.1	6.0	80.0	14.0	80.0	11.0	9.0
14.	8.9	7.7	1.2	6.0	78.0	16.0	69.0	28.0	13.0
15.	7.8	7.4	0.4	9.0	76.0	15.0	75.0	13.0	12.0
16.	7.9	7.6	0.3	9.0	77.0	14.0	82.0	16.0	12.0
17.	7.4	7.2	0.2	5.0	60.0	35.0	84.0	15.0	11.0
18.	7.9	7.2	0.7	8.0	69.0	13.0	76.0	8.0	16.0
19.	7.9	6.9	1.0	9.0	71.0	20.0	65.0	20.0	15.0
20.	7.5	7.4	0.1	10.0	78.0	12.0	75.0	15.01	10.0

Conclusion

From the discussion it can be concluded that ecosystem of low biomass, small structure and low productivity do not support the construction of mounds therefore the region had no mounds before the plantation. The activity of constructing mounds is seasonal, i.e., in rainy season construction is absent and in the month of high temperature (above 30°C) repairing work is carried out. If there is no obstacle for construction then generally mounds are in north-south direction and north side is slightly larger than south which indicate the definite magnitude of mound. All mounds in the study area are of Bissel type.

Generally, mounds have more pH than surrounding soil which is related to the location of mound. The mounds near the drainage system and in open areas have more pH due to the capillary action and evaporation. The clay and silt content of mound soil is always more as compared to surrounding soil, which indicate the ability of termites to mobilise particles and forming a prominent micro-relief feature.

References

Cowan, J.A.; Humphreys G.S.; Mitchell P.B. and Murphy C.L. (1985): An assessment of Pedoturbation by two species of mound-building ants, Componotus intrepidus (Kirby) and Iridomyrmev purpureus (Smith). *Australian Journal of Soil Research*; 22; 95-107.

Darlington, J.P.E.C. (1984): Two types of mound building by the termite Macro-termites Subyalinus in Kenya. Insect Science Applications; 5; 481-92.

Ehrich, H.L. (1981): *Geomicrobiology*. Marcel Dekker, New York.

Harris, W.A. (1961): *Termites: Their Recognition and Control*. Longmans, London.

Krumbein, W.E. (ed.) (1983): *Microbial Geochemistry*. Blackwell Scientific Publishers, Oxford.

Krumbein, W.E. (ed.) (1978): Environmental Biogeochemistry and Geomorphology. 3 vols; Ann Arbor Science Publisher, Ann Arbor.

Maeterlinck, M. (1927): Life of White Ant; Allen and Unwin, London.

Mohindra, P. Mukerji K.G. (1982): Fungal ecology of termite mounds. Revue d'ecologie et de biologie du sol, 19, 351-61.

Peter, Ferb and the Editors of Life (1964): The Insects, Times Life International, (Netherland) N.V.

Skaife, S.H. (1955): *Dwellers in Darkness*, Longmans.

UNESCO\UNEP\FAO (1979): Tropical grazing land ecosystems; UNESCO, Paris.

Viles, H.V. (ed.) (1988): *Biogeomorphology*, Basil Blackwell, Oxford.

9

Gully Erosion and its Implications on Landuse: A Case Study of Dumka Block, Dumka District, Jharkhand

V.C. Jha and Saswati Kapat

Accelerated soil erosion and land degradation, caused by natural phenomena, indiscriminate destruction of vegetation and changes in landuse are the burning issues of today.

More than 50% of the total area of India is affected deleteriously by land degradation resulting from soil erosion (Sehgal and Abrol, 1994; Mandal, 1996; Biswas, 1999), major contributing factor of which is gully networks and other forms of water induced soil erosion. They are still at work to convert usable land into degraded/wasteland. On the other hand, the growing demand of India's ever increasing population on land requires continuous effort for optimum and rational utilisation of land in consistent with sustainable development.

For this purpose, suitable planning measures should be devised in such a manner so that even existing waste lands may be brought under plough or other uses. In fact, wasteland category of land can be made suitable for use of mankind by proper treatment under scientific management (Jha, 1987, p. 231). But an essential prerequisite to any planning for such management is the identification of the productive or usable areas, that are likely to be affected by the process of degradation and the areas that have already been damaged (Mensching, 1982, p. 94). In fact, erosion and land degradation are the outcome of physical and socio-economic process (Perez, 1998, p. 294).

Therefore, comprehensive geomorphic study with special reference to morphodynamic processes and associated assemblages of landforms is essential to ascertain the nature of land in terms of intensity of degradation and usability in promoting socio-economic development of the area.

This paper attempts to determine the nature, extent and distribution of gully networks particularly in association with climatic, pedogenic, geomorphic and anthropogenic aspects and to define the implication of gully development in terms of the extent of gullies at work in to usefully potential and even cultivated/forest land. In addition of this, the classification of the gully affected land according to the nature of gully erosion has been made for devising remedial measures of the study area.

Study Material and Methods

In the present study, visual interpretation of satellite imagery (LS-5, MSS FCC, April, 1990) for the demarcating and grouping of gully affected areas in different geomorphic domains has been done. Toposheets (Number 72/P, 72 $^P/_3$, 72 $^P/_7$ and 72 $^P/_8$ scale 1: 25,000 and 1: 50,000) have also been studied in detail for deriving pertinent informations about relief, gradient and drainage which govern the tractionability of the terrain.

Thornthwaite-Mather's book keeping procedure of computation for water balance (used by V.P. Subrahmanyam) has been adopted to determine soil moisture storage, soil moisture surplus, moisture index and runoff. This computation has been done on the basis of data of monthly temperature and precipitation over a period of 1990-1999 (data source, IMD-Delhi and Birsha agri. University, Dumka, Jharkhand).

The general equation derived for it as follows:

$$PET = 1.6 \ (10t/I)^a$$

where,

PET = Potential Evapotranspiration on monthly basis;

I = Annual Heat Index which is the sum of 12 monthly indices i,

i = $(t/5)^{1.514}$;

a = 6.75 x 10^{-7} I^3 - 7.7 x 10^{-5} I^2 + 1.792 x 10^{-2} I + 0.49239. (Thornthwaite's Method*)

(ii) Im = (Ih - Ia)

Im = Moisture Index, Ih = Humidity Index, Ia = Aridity Index (Thornthwaite and Mather's Method 1955*)

(iii) Runoff: Runoff has been determined by the surplus precipitation (from total to PET) reduced to 0.65 of its computed value. (Subrahmanyam 1980, p. 88).

The monthly and seasonal concentration, duration and intensity of rainfall and total number of rainy days have also been taken into account to infer its erodibility. Actually all these components apart from geomorphic, pedogenic, vegetative and anthropogenic factors play important role in gully formation (A gully refers to a trench worn in the earth by the action of water after heavy downpour).

Five gullies with 20 samples sites on contrasting terrain have been selected to note their erosional characteristics in relation to their implication on landuse (Table 1, Figure 2). Field survey was spread over the pre(March-May), South-West (June- September), retreating (October-November) and also winter monsoon seasons for recording measurement with respect to gully scars/gully affected parts. The S/L ratio, i.e. ratio between the volume removed by the side-wall erosion(s) and the channel erosion (L) of gully according to Blong's method (1982) has been computed on the basis of the data obtained from district soil department (Dumka). Finally, attitude of gullies towards existing landuse pattern and distribution in correlation with above mentioned attributes has been determined.

Study Area with its General Informations

Location: Dumka block, (District Dumka, Jharkhand Pradesh, India) belonging to the south western part of the Rajmahal

Figure 1
Location

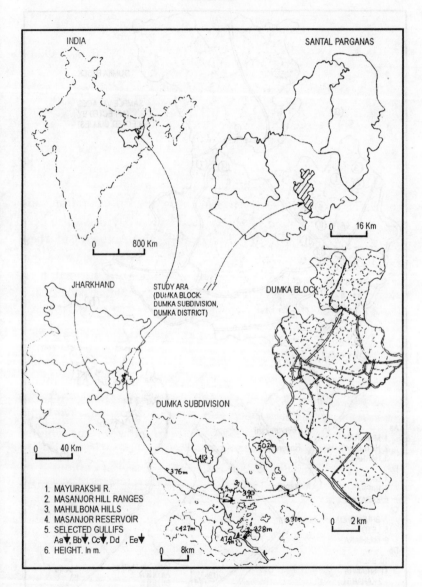

INDIA

SANTAL PARGANAS

0 ___ 800 Km

0 ___ 16 Km

JHARKHAND

STUDY ARA
(DUMKA BLOCK:
DUMKA SUBDIVISION,
DUMKA DISTRICT)

DUMKA BLOCK

0 ___ 40 Km

DUMKA SUBDIVISION

1. MAYURAKSHI R.
2. MASANJOR HILL RANGES
3. MAHULBONA HILLS
4. MASANJOR RESERVOIR
5. SELECTED GULLIFS
   Aa▼, Bb▼, Cc▼, Dd , Ee▼
6. HEIGHT. In m.

0 ___ 8km

0 ___ 2 km

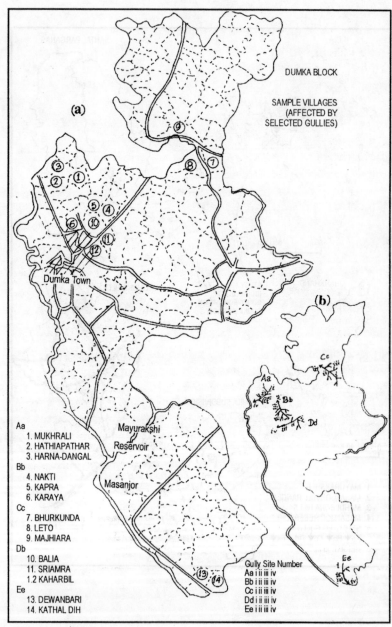

Figure 2
Sample Villages

highland lies between 24°3'N and 24°27'N and 87°13' and 87°27'E with an area of 378.588 km². (Figure 1).

Relief: The general elevation varies between 100m (down the Mosanjor reservoir) and 393m (Mohulbona hills). Most of the area is dominated by gentle undulating planation surface (85.21% of the Dumka block) with two parallel hill ranges and a prominent isolated hill to the north-east of Mosanjor reservoir respectively.

Geology: The area is predominantly composed of gneissic rock with some occurrence of gabbro, dolerite, hornblend schist, etc.

Climate: The area belongs to the monsoon type of climate. According to moisture index, done on the basis of Thornthwaite - Mather's method, it is dry sub-humid in nature. The average monthly temperature ranges from 15.17°C (January) to 30.15°C (May); average monthly rainfall from 2.62mm (December) to 339.4mm (July) and average monthly rainfall intensity from 1.80mm (December) to 17.98mm (August).

Morphogenetic region: It falls under moderate morphogenetic region.

Soil: The soilscape of the area is constituted by red-yellow and lateritic soil (Venkataraman and Krishnan, 1992, p. 479) under the soil series of Dumka, Pusaro. Lachimpur, Sarua, Karaya, Hathiapathar and Kanchanpur. Soils vary in colour from yellowish brown to strong brown. Soil textures of surface and subsurface horizons are of gravelly sandy, sandy clay loam and clay loam-clay respectively. Sesquioxide is common in the B horizon. (Birsha agri. Research Centre 1994).

The infiltration rate of different types of loam soil ranges from 8 mm/h to 15 mm/h and pH value from 5.2 to 7.4.

Drainage: It is mostly drained by the Mourakshi river and its tributaries. The Brahmani river with its tributaries flows over the northeastern part of the area.

Vegetation: Vegetation is sparse except on hill slope which bear forest cover. Vegetation is mostly deciduous and shrubby/scrubby in character.

Morphometric attributes (1 sq. km. grid): The area is mostly dominated by low absolute relief (AR), relative relief (RR), dissection index (DI), average slope (AS), drainage frequency (DF), and density (DD). The ranges of these morphometric parameters are as follows:

AR: 100m - 393m; RR: 11m -233m; DI: .03 - .593; AS: .50° - 14.12°; DF: 0-7/km^2 and DD: 0 to 5.6 km/km^2.

Gullied Area: According to visual interpretation of satellite imagery the gully affected area, identified by yellow and yellowish white tone, accounts for 26.4% of the Dumka block.

Landuse: The classes of landuse of the area are as follows: net cultivated area - 42.34% of the block; current fallows - 3.74%; area under forest - 11.40%; area not available for cultivation - 18.14%; other uncultivated area - 9.22%, area under waste land - 15.16% (Report of Lead bank, Dumka and statistical division, Dumka 1991-1998) Figure 3(b).

Village and Town: It has 276 villages (19 uninhabited) and one town-Dumka town (1991).

Result and Analysis

Rain, Run-Off, Gully Erosion

High concentration of rainfall in the study area, reckoned from monthly/seasonal total rainfall occurring in the south-west monsoon season (June-September) has been given much emphasis as in this season about 83.5% of the yearly total is received. Moreover, in the month of August the highest amount of hourly rainfall (49.5 mm/h) is recorded; the maximum value of monthly rainfall intensity (27.55 mm) is also noted in this month. Therefore, the average seasonal rainfall intensity (19.79 mm) tends to be high in south-west monsoon season. In the pre, post/retreating and winter monsoon season it is very low - 7.76 mm, 10.23 mm and 3.37 mm respectively (Table 3). Under the prevailing dry sub-humid climatic condition (Im = -12.31%), each month from July to September in south-west monsoon season having higher rainfall than potential evapotranspiration yields water surplus which amounts to, 138 mm, 156 mm, 136 and respectively. The water surplus in each of these months generates runoff and it is estimated at 167 mm, 133 mm and 123 mm and

Figure 3

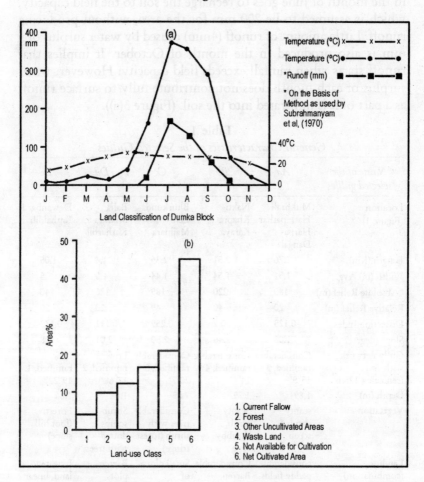

(a)

Temperature (°C) x—— x —— x
Temperature (°C)●—— ● —— ●
*Runoff (mm) ■—— ■ —— ■

* On the Basis of
Method as used by
Subrahmanyam
et al, (1970)

Land Classification of Dumka Block

(b)

1. Current Fallow
2. Forest
3. Other Uncultivated Areas
4. Waste Land
5. Not Available for Cultivation
6. Net Cultivated Area

Land-use Class

respectively (Table 4). It would be appropriate to mention that water excess (because of more specification than the reduced PET) in the month of June goes to recharge the soil to the field capacity, which is assumed to be 200 mm for the area, without producing runoff. Little amount of runoff (4mm) caused by water surplus of 3 mm is also produced in the month of October. It implies that runoff starts when rainfall exceeds field capacity. However, water surplus of each month does not contribute fully to surface runoff as a part of it is infiltrated into the soil. (Figure 3(a)).

Table 1
General Characteristics of the Selected Gullies

Name of the selected gullies	Aa	Bb	Cc	Dd	Ee
Location (Figure 1)	Mukhrali Hathipathar Harna-Dangal	Nakti-Khapra-Karaya	Bhurkunda-Leto-Majhiara	Balia Sriamra-Karharbil	Dewanbar I Kathaldih
Length (Km)	2.76	2.51	2.35	1.4	1.05
Width (m) Avg.	1.51	1.54	1.44	1.0	1.5
*Absolute Relief (m)	180	220	189	180	140
Relative Relief (m)	20	40	49	20	20
Dissection Index	0.125	0.2	0.259	0.111	0.172
Slope (degree)	1.03	2.46	2.50	2.01	1.8
Gully network with Avg. frequency 125m^2	Considerably ramified, 3 15°9'	Considerably ramified, 4 14°6'	Considerably ramified, 5 20°7'	Low ramified, 2 17°68'	Low ramified, 1 19°22'
Depth (m)	1.891	1.79	2.25	1.86	1.66
Vegetation	scanty shrub & scrub	Scanty scrub, shrubby type	Considerable trees with shrub (forest fringe)	Shrub-scrubs dotted with trees	Forest (foot hill slope)
Landuse (combination)	considerable paddy fields, Barren plots, fallow land, pasture, compact settlement	Considerable Barren, fallow and few cultivated land Road (State highway), sparse/linear settlement	Small plots of cultivable, considerable uncultivable land, road & sparse settlement	Barren plots, uncultivable land, deserted/spa rse settlement	Irrigated land, linear settlement, close to reservoir and canal

* Values of relief morphometric attributes of terrain bearing gullies, derived from grids of 1 sq. km. and 125 m^2 only in case of gully network.

Table 2
Forms and Processes of Gullies According to Field Observation (1990-1999)

Forms:	Pinnacles, near vertical or quite steeply sloping fluted side wall, acute bend at the gully head wall, scoured fluted small residual buttress. Scar of detached slab from head and side wall, holes, ephemeral small pipes, broad flat gullied bottom or narrow bottom, Partial collapsing features etc. - all these features existing in all the gullies to small and large extent.
Processes:	Fluting, scouring, wall failure, piping, basal sapping flattening of inter-gully area, etc.

Table 3
Temperature, Rainfall, Rainy days and Rain Intensity (1990-1999)

Season	Temperature (°C)			Amount of Rainfall (mm)		Highest % of rainfall is S.W. monsoon season	Rainy days	Intensity of Rainfall (mm/month)	
	Monthly Avg.*		*Avg. (seasonal)	Max (Monthly total in different seasons)	Avg. (seasonal)		Average	Maximum in season	Avg. in season
	Max	Min							
Pre-Monsoon season	33.23 (May)	22.9 (Mar.)	29.61	76 (Mar.)	28.36		3	18.33	7.76
S.W. Monsoon season	30.2 (Jun.)	23.85 (Sept.)	27.58	242.2 (July)	299.32	83.5%	18	27.55	19.79
Retreating-monsoon season	29.33 (Oct.)	16.41 (Nov.)	23.5	173 (Oct.)	48.56		6	19.22	10.23
Winter season	19.5 (Dec.)	13.95 (Jan.)	17.45	31 (Feb.)	7.5		2	9.33	3.37
Annual Avg+	24.32			945.86					

Annual Avg+ - For 10 years.
*Avg - Average for the period of 10 years (seasonal).
Monthly Avg* - Calculated for the month of each year, Max & Min monthly Avg. temperature obtained in particular years of the decade. Maximum monthly total rainfall and intensity of rainfall obtained in particular years of the decade.

Table 4

Water Balance Characteristics (1990-1999)

	Range of Avg. monthly Potential Evapotranspiration (mm)	Range of average monthly soil water storage (mm)	Range of average water deficit (mm)	Range of avg. W. Surplus & Avg. water surplus in S.W. and retreating monsoon season (mm)	Avg. run off (Computed) in S.W. & retreating monsoon season (mm)	Moisture Index	Climatic Type
Max	182 (May)	200 (S.W. monsoon)	133 (May)	143 (August)	167 (July); 133 (August)	-12.31%	Dry Sub-humid
Min	23 (January)	12 (May)	0 (S.W. monsoon)	0 (Most of the month); Avg. in months of S.W.	123 (September); 3 (October)		
	Pre Monsoon - 148, S.W. Monsoon - 155, Post Monsoon - 78, Winter - 23	(Soil water storage reaching field capacity only in S.W. monsoon season)	(Only S.W. Monsoon Season - devoid of water deficit)	Monsoon season - 156 (July), 138 (August), 136 (September), 4 (October)			

field capacity 200 mm.
Run off - Calculated from reduced PET by 65% - not shown in this table.

Table 5

Seasonal Average Head and Side Wall Erosion of Different Selected Sites at Interval of Two Years (1990-1999)

Pre-Monsoon Season				S.W. Monsoon Season			
Head wall		Side wall		Head wall		Side wall	
Length (m)	Width (m)	Length (m)	Width (m)	Length (m)	Width (m)	Length (m)	Width (m)
.20	.18	.16	.12	.78	.59	1.60	1.08
Post Monsoon Season				Winter			
Head wall		Side wall		Head wall		Side wall	
Length (m)	Width (m)	Length (m)	Width (m)	Length (m)	Width (m)	Length (m)	Width (m)
.65	.48	.67	.56	.04	.02	.03	.02

According to Table 6, a considerable formation of gully scars on head and side wall ranging from 0.56 meter to 1.93 meter in length; 0.35 to 1.5 meter in width; 59.5 degree to 72.5 degree in slope are noticed in South-West monsoon season when it receives maximum amount of rainfall with high intensity. The pre and post monsoon seasons particularly the former, with short lived torrential downpour, are efficacious in gully erosion. It is noted specially in South-West monsoon season that side wall collapse, resulting from mass movement, is caused by its basal soil saturation/abrasion by water within the trench. Again water that infiltrates into the horizontal soil surface of the head cuts and moves downwards under gravity to the base of the vertical wall, and tends to cast off the supporting base of the gully head. In the study area, sandy clay loam, which is dominating in the surface horizon, is thus responsible for weakening the supporting basal foundation of the vertical headcut. Generally this phenomenon is found to occur in rainy or post monsoon season. Such investigation supports the statement of (Ireland Sharpe and Eargle

Table 6

Characteristics of Scars Appeared in Different Seasons During Study Period (1990-1999)

Name of Gullies	Site No.	Date of Observation	Rainfall (MM)		Season	Gully-Scar (Local)				Remarks
			T.M.*	H.R.*		Wall	Length (m)	Width (m)	Slope (Degree)	
Aa	Aai	7.09.96	895.6	1.32 (on 6th August)	S.W. Monsoon	Head	.88	.41	59.5	Upper part of head wall collapsed
	Aaii	7.09.96				Side	1.50	.45	71.01	Widely serrated side wall bearing small tunnels and holes; its collapsed part becoming steep
Bb	Bbii	30.06.93	229.6	66.4 (28th June)	S.W. Monsoon	Side	1.93	1.55	72.05	Arched scar appeared at side wall's bottom removing a considerable mass
Cc	ccii	15.10.94	110.4	58 (10th October)	Post Monsoon	Head	1.26	.92	71.01	Upper part of head wall collapsed; crescent scar
	Ccii	18.10.94	110.4		S.W. Monsoon	Side	.94	.64	70.03	Hole to be collapsed observed near side wall. Side wall with a tree collapsed
Dd	Ddi	28.09.92	954.59	140.5 (20th September)	S.W. Monsoon	Head	.56	.35	65.03	Basal scar with irregular head
	Ddii					Side	1.01	.805	50.41	Upper part of projected side wall with collapsed pinnacles, with head broken down
Ee	Ffii	30.05.98	129.5	39.6 (10th March)	Pre-Monsoon	Head	.68	.46	62.2	Crescent head wall scar

T.M. * Total amount of rainfall upto observation date of the season.

H.R. * Day of Highest Rainfall during that season in a particular year.

1939; Bradfort, 1977, Hook, 1979). Besides this, effect of rain splash and collapsing of small tunnels are found frequently, particularly in the gullies namely Aa, Bb, Cc. Similarly Table 5 reveals same picture with regard to frequent appearance of gully scar in south west monsoon season.

Hence south-west monsoon season brings about greater probability for gully erosion both headward and sideways than other seasons, particularly in winter season. In fact, there is a significant correlation (r + 0.58) between the number of rainy days collectively causing high concentration of seasonal rainfall and the number of gully scars.

Gully Erosion and Soil

With gentle slope and appreciably sparse vegetation, the area is predominated by sandy clay loam texture in the surface horizon and clay loam/clay with ample sesquioxides (aluminium and ferric oxides) in the sub-surface horizon. The structure varies from weak to moderate sub-angular blocky in the surface horizon with soil consistency varying from friable and non sticky to firm and sticky, while in sub-surface horizon, it is moderate to strong sub-angular blocky with extremely firm, sticky and plastic soil consistency (Table 7).

Basically such soil texture and structure provide the area characteristics for surface and sub-surface soil erodibility and is prone to piping and guly erosion. Considerable infiltration, which is found to be 12 mm/h or more in the A horizon, especially in the sites (such as Balia, west of Nakti, etc.) of sandy and gravelly loam and low permeability (less than 7mm/h) in clay enriched illuviated B horizon (as noted in the Table 7) cause the development of pipes most of which are ephemeral, active only in times with high concentration of rain fall. The presence of holes and tunnels on the surface or gully walls found in the fields proves their existence. Ultimately the collapse of the roofs of pipes creates a gully (Wild 1993, p. 240; Jones, 1997, p. 88-89).

Table 7
Soil Characteristics of Gullied Land

Soil Type	Horizon	Soil According to Texture	Dominating Texture	Gravel %	Sand %	Silt %	Clay %	Dominating Structure & Soil Consistence	Drainage
Red-Yellow & Lateritic	Surface	Sandy Loam	Sandy-clay Loam		65	20	15	Moderate subangular Blocky, Firm & Sticky	Moderate of well
		Gravelly Sandy Loam		25	40 & more	20	15		
		Sandy Clay Loam			20 & more	20	35		
		Silty Clay Loam				40 & more	40		
		Clay Loam			10	45	45		
	Sub Surface	Gravelly Sandy Clay Loam & Sandy Clay Loam	Clay Loam & Clay	5	40 & more	20	35	Sub-Angular Blocky, Extremely Firmly Sticky Plastic	Moderate to poor (Poor-dominating)
		Clay Loam & Clay with Ferruginous Concretion			10	45	45 & more		

Source: Birsha Agri University, Bihar.

Linear and Side Wall Erosion of Gully with S/L Ratio Index

The estimated S/L ratio between volume of eroded materials resulting from channel/linear (L) and side wall erosion (S) at different parts of the gully reveals that the side wall erosion is in general more dominant in the lower part and also middle part of the gully while the linear one in the upper part. It is displayed almost in all cases of gullies of the area, as corroborated by the following table of classified S/L ratio. In fact low S/L ratio (less than 1), suggests acceleration of head wall erosion generally forming narrow trench while high value of it (>5 or around 5) does in the side wall for its widening. In fact the former two types of S/L ratio mentioned in Table 8 are noticed in each selected gully. However, gullies namely Aa, Bb and Cc receive distinctly S/L ratio >5 in their middle or lower part.

Table 8

S/L Ratio and Percent of Avg. Volume of Eroded Materials from the Gully Wall: (Since its Initiation)

S/L ratio	Avg. % of the volume of eroded materials of channel process	Avg. % of the volume of eroded materials from side wall process	Characteristics of S/L ratio in term of processes	Part of the gully
less than 1	68.20	31.8	Channel erosion with active head cut	The upper part, nearly up to .89m from the head with narrow trench
1 to 5	42.5	57.5	1 S/L ratio means channel and side wall erosion equal. Appreciable side wall erosion	Middle and lower part with appreciable wide bottom
More than 5	29.68	70.32	More dominant side wall erosion	(occasionally in the middle part)

A close perusal of Tables 2, 5, 6 and 8 demonstrates distinct dominance of side wall erosion which tends to flatten the gully ridge crest and inter-gully sites of the study area which is generally characterised by low magnitude of terrain attributes (Table 1).

Table 9

% of Gully Affected Area in Terms of Landuse

Name of sample villages affected by gully erosion	Area (ha)	Forest	Cultivated area	Fallow land	Irrigated part	Non-agricultural gullied land	Settlement area	Gullied waste land
Mukhrali	81.28							4.21
Hathia pathar	180.18		4.81	1.58	0.82		1.04	3.06
Harana dangal	106.68						1.31	4.90
Nakti	233.98		2.91				2.86	10.75
Khapra	75.11			1.38				5.32
Karaya	175.74							1.62
Bhurkunda	105.62	1.62	3.51			1.03	0.56	6.24
Leto	180.10	0.91						6.15
Majhiara	193.14	1.21						6.76
Balia	66.48							6.16
Sriamra	120.49		1.59			0.48	1.69	4.10
Kaharbil	125.69			0.61				1.38
Dewanbari	144.75	0.53			1.26	1.58		1.09
Kathaldih	177.12	0.89			0.75	0.48		2.85
Total	1966.36	5.16%	12.82%	3.57%	2.83%	3.57%	7.46%	64.59%

Gully Erosion and Landuse

It is evident from the field investigation that all the landuse classes of the Dumka block, are victim of gully erosion. The Table 9 reveals that the majority of the area (64.59%) of the sample villages comprises gullied waste land which is in the gradual increase when potential use of land, available in a maximum possible way in the study area, is needed to remove the 'backwardness' of this block as well as the entire district. It is to be mentioned that the Planning Commission declared it as backward district (Report of Lead Bank, Dumka, 1991-1998). Particularly gullies namely Aa, Bb and Cc characterised by comparatively larger size net work, give rise to most gullied waste land surface. Surface soil erosion also plays an important role for worsening the condition of the land surface of some sites. About 12.82% of the total cultivated/cultivable land

Table 10
Classification of Gully Affected Area

Type of gully affected area	% of areal coverage of gullied area	Name of village	Characteristics	Pattern of Landuse
Highly affected gullied area	28.20	Hathia Pathar Nakti, Balia, Bhurkunda etc.	Apprecibly dissected/rough gullied terrain with highly ramified gullies, gully frequency 4 and more/125m²., depth ranging from 2.3m-3.4m more: width, more than 1.9, considerable occurrence of gully erosion scar, occurrence of much pinnacles	Waste land Fallow land Non agricultural land
Moderately affected gullied Area	37.62	Mukhrali, Khapra, Leto etc.	Gullied terrain with the frequency of gully 4/125m². Depth ranging 1.5m to 2.5 m & more, considerably wide flat bottom, occurrence of side wall erosion scar.	Fringe of forest, cultivated/cultivable land, Close to settlement, Agricultural Non agricultural land (being used)
Least affected gullied area	33.18	Kaharbil, Karaya, Kathaldih etc.	Gullied terrain with the frequency of gully-2 or less, Depth -less than 2m, small side & head wall scar.	Forest fringe, Settlement, Arable land, Canal, Road.

have been eaten by gullies. The dismal condition of the cultivated, and cultivable land has been aggravated more by the addition of 14.6% (3.57% fallow land, 7.46% land close to settlement and 3.57% non agricultural land) gullied land caused mainly by anthropogenic factors (viz. vegetation cutting for fuel and fodder, rampant illegal rock extraction, unplanned non-metalled road and other constructional works etc.). The villages under gullies——Aa, Bb, Dd are totally deprived of forest cover; only degenerated forest being dwindled more by tree cutting remains in those of Cc and Ee and it accounts for only 3.74% and 1.42 respectively i.e. only 5.16% of the total area of the selected villages. It is evident from the ramification of the gully area and the trend of the degeneration of forest cover that the land resources have been deteriorated to a great extent over a period of time (1990-1999) in the sample area or the whole Dumka block. In addition to this, canal irrigated area (in the vicinity of canal) which is very limited (2.83% of the total area of villages) in this block is also being encroached by gullies. In the context of above views, gullied land has been categorised in to three groups such as highly (28.20%), moderately (37.62%) and least (33.15%) affected gullied areas (Table 10). In fact the group of highly affected area is still at the stage of vulnerable soil erosion. (Figure 4).

Change in the uses of the arable land is also responsible for erosion and land degradation apart from the cause as reflected in the Report of Dumka, statistical Division - 'the arable land of Dumka was estimated to 64.2% in 1961-1971. At present it has been reduced to 42.34% in view of the increase of non-agricultural use of land.' Progressional land degradation on the one hand and decline of arable land on the other reflect gradual decrease in the area of the family holdings. Small holding with low fertility of soil and high dependence on rainfall for cultivation do not provide sufficient earning to the cultivators to raise their life standard to an expected level. Many families are still living below poverty line. It is surprising to note that Balia and other nine villages, affected heavily by gullies which have converted a great part in to gullied wasteland, is totally deserted. This abandonment has set an example of poor economic condition prevailing among most of the families of the area. Such poor economic condition is still unabated in the area, though development programme in many aspects has been carrying on for decades under IRDP and NABARD. Recently these agencies have given emphasis upon collective effort on reclamation and proper utilisation of waste land, restoration and conservation of village resources and cautious application of modern technique to increase the yield of agricultural produces vis-a-vis income of the inhabitants with a view of environmental and sustainable development by bringing balance between their short term and long term needs.

Figure 4
Gully Affected Area

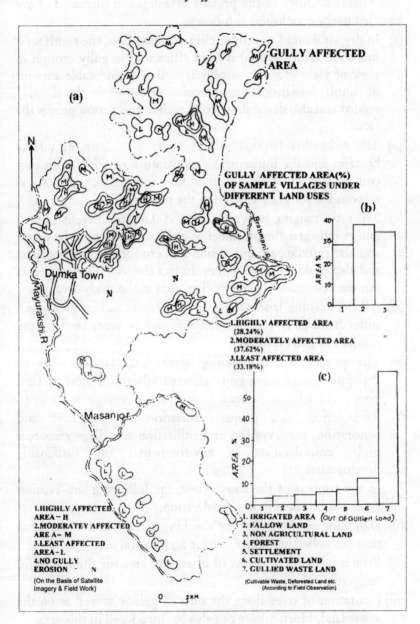

(a)

GULLY AFFECTED AREA

N

Dumka Town

Mayurakshi R.

Brahmani R.

Masanjor

N

N

N

GULLY AFFECTED AREA(%)
OF SAMPLE VILLAGES UNDER
DIFFERENT LAND USES

(b)

AREA %

1.HIGHLY AFFECTED AREA (28.24%)
2.MODERATELY AFFECTED AREA (37.62%)
3.LEAST AFFECTED AREA (33.18%)

(c)

AREA %

1. IRRIGATED AREA (OUT OF GULLIED Land)
2. FALLOW LAND
3. NON AGRICULTURAL LAND
4. FOREST
5. SETTLEMENT
6. CULTIVATED LAND
7. GULLIED WASTE LAND

1.HIGHLY AFFECTED AREA – H
2.MODERATEY AFFECTED AREA – M
3.LEAST AFFECTED AREA – L
4.NO GULLY EROSION – N

(On the Basis of Satellite Imagery & Field Work)

(Cultivable Waste, Deforested Land etc.
(According to Field Observation)

0   2 KM

Conclusion

The salient findings of the present investigation discussed above have led to the conclusion as follows:

(a) In dry subhumid monsoon climatic condition, the south west monsoon (rainy season) is most efficacious in gully erosion in view of yield of water surplus as well as considerable amount of runoff. Initiation of new erosion processes on the severely eroded and also destabilised gully wall in this season proves the fact.

(b) The red-yellow-lateritic soil of sandy clay loam in surface horizon and the impermeable substratum enriched with clay content, sesquioxide, aluminium and ferric oxides, is responsible for gully erosion in the pedosphere.

(c) S/L ratio ranging between 0.35 and 0.61 in different sites of gullies indicates the prominence of linear erosion in the upper and also middle part and of side wall erosion in the lower part and also middle part of gullies. In fact the over all view points out the dominance of side wall erosion in the study area.

(d) All the existing landuse classes even forest and irrigated land suffer from gully erosion. The majority of waste land are the product of gully erosion.

(e) The process of abandoning agricultural land is due to depopulation in some gully affected villages in view of very poor income from the land.

(f) Reclamation and proper utilisation of wasteland and restoration, conservation and utilisation of village resources under consideration of environmental and sustainable development.

In the context of the above view, the following anti erosion measures can be taken in to consideration:

(i) Damming of the gullies for checking erosion and at the same time for collecting rain water for agricultural use.

(ii) Proper land management of intergully area for the successful cultivation of crops.

(iii) Plantation of trees along the sides of gullies as well as on the waste land. Horticulture can also be introduced in this area.

Acknowledgement

I gratefully acknowledge Shri. G. Kapat (Retd. formerly HOD Geography, Visva-Bharati) and Prof. V.C. Jha (Visva-Bharati, W. Bengal) for continuous encouragement, their critical comments, and help in carrying out this work. I also thank to P.K. Shaha and others, Birsha Agricultural University, Dumka for providing relevant data and reports.

References

Biswas, et. al. (1999): Prioritisation of sub watersheds based on morphometric analysis of drainage basin, *Photonirvachak*; Jour. of ISRS, Vol. 27, No. 3, p. 155.

Blong, R.J. (1982) The role of side wall processes in gully development, some N.S.W. examples; Earth Surface Process and Landforms, Vol. 7, pp. 381-385.

Joshi, V. and Kale, V. (1995): The contribution of side wall erosion in the gully development, *Indian Geomorphology and Resource Management*, Rawat Publications, Vol. 1, pp. 43-54.

Jones., J.A.A. (1997): Subsurface flow and subsurface erosion; *Process and Form in Geomorphology*, pp. 88-89.

Leopold, B.L. et. al. (1964) *Fluvial Processes in Geomorphology*, pp. 446-447.

Mandal, C. and Mandal, D.K. (1996) Qualitative assessment of soil erosion from soil survey data: A case study of Nagpur district, *Geographical Review of India*, Vol. 58, No. 1, p. 294.

Menshing, H. (1982): Applied geomorphology example of work in the tropics and the subtropics; *Applied Geography and Development*, 19 Vol, p. 94.

Garcia Perez. J.D.: Geo OKO Dynamik XIX-1998: The use of pair wise photographs for soil and water conservation planning: a case study of Guadalazara, Spain, p. 293.

Report of Lead Bank and District Statistical Department 1991-1998.

Soil Survey Report (1994), Birsha Agricultural University, Dumka.

Sehgal, et. al. (1994): Red and lateritic soil of India: Resource appraisal and management, NBSS, pp. 5-7.

Subrahmanyam, V.P. (1982): *Water Balance and Its Application*, Andhra Univ. Press, Waltair, pp. 13-89.

Thronthwaite, C.W. and Mather, J.R. Instruction and tables for computing potential evapotranspiration and water balance, Publication in Climatology Centerton, Vol. 10, 3 (1957).

Venkataraman, S. and Krishnan, A.: *Crops and Weather*, Publications and Information Division, Indian Council of Agricultural Research, Pusa, New Delhi, pp. 478-479.

Wild, Alan (1993): *Soil and the environment: An introduction*, Cambridge Univ. Press, pp. 19-240.

10

Land Degradation Issues in the Coal Mining Region of Raniganj

Kuntala Lahiri-Dutt and Prasanta K. Jana

Dimensions of Degradation

'Degradation' is defined as a qualitative deterioration of the production potential of any element of the environment. Land, along with the thin layer of soil cover on it, is the most valuable element supporting human, animal and plant life. It is also the element which continuously negotiates between the cultural and physical components of ecosystems (Brookfield and Padoch, 1994).

Agenda 21 argued for both capacity-building and the promotion of ecologically and culturally sensitive schemes of land restoration. 'Capacity-building' means expanding the vernacular science of local knowledge and tradition, and integrating it with the western science of soil and water management, monitoring of rainfall and groundwater movements, and analysing changes in soil fertility following different treatments. Thus, land itself becomes a part of the cultural milieu, as local users get precedence over outside scientists in the maintenance of the quality of land.

Many authors do not make any differences between 'erosion' and 'degradation' and 'soil' and 'land'. They use them interchangeably in the literature. However, there are some important differences. Stocking (2000) recently gave clear-cut definition about 'soil erosion', 'soil degradation' and 'land degradation'. 'Soil erosion' is defined as one of the main processes of degradation, which consists of physical detachment of soil particles by wind and water and their transport to other parts of the landscape, to rivers and the sea. But, 'soil degradation' is defined as a decrease in soil quality as measured by changes in soil properties and processes, and the consequent decline in productivity in terms of immediate and future production (Table 1).

Experts (Stocking, 2000) identify six processes of soil degradation:

- *Water erosion*: splash, sheet and gully erosion, as well as mass movements such as landslides.
- *Wind erosion*: the removal and deposition of soil by wind.
- *Excess of salts*: processes of the accumulation of salt in the soil solution (salinisation) and the increase of exchangeable sodium on the cation exchange of soil colloids (sodication or alkalinisation).
- *Chemical degradation*: a variety of processes related to the leaching of bases and essential nutrients and the building up of toxic elements; pH related problems such as aluminium toxicity and P-fixation are also included.
- *Physical degradation*: an adverse change in properties such as porosity, permeability, bulk density and structural stability; often related to a decrease in infiltration capacity and plant-water deficiency.
- *Biological degradation*: increase in the rate of mineralisation of humus without the replenishment of organic matters.

Seen from this perspective, 'land degradation' is a composite term signifying the temporary or permanent decline in the productive capacity of the land. It is the aggregate diminution of the productive potential of the land including its major uses (agriculture, forestry, urban landuse, industrial landuse, mining landuse) and its value as an economic resource.

Enters (1997), however, defines land degradation as reduction in the capacity of the land to produce goods and services for the humankind. He indicates it as 'more than just a physical or environmental process. Ultimately, it is a social problem with economic costs attached as it consumes the product of labour and capital inputs into production' (Enters, 1997).

GlASOD (Global Assessment of Land Degradation, referred in Oldemann *et al.*, 1990) was the most important study on land degradation to generate analysis on a continental scale. However, micro-level studies of land degradation due to specific causes, and the human dimensions of such degradation are of great significance as they help to bring out the local issues.

Mining and Land Degradation

Mining leads to the biophysical damage to soils, accumulates important on-site and off-site economic costs, and results in the displacement of local communities. The degree of degradation, cost and disruption varies according to soil type, landuse and the process of degradation. There is, however, much less mention of mining in the land degradation literature as a significant 'cause'.

Table 1

Degradation Status of Global Land in Crops, Permanent Pasture, Forest and Woodlands

Degradation Category	Amount of Land Affected (million hectare)	Lost Production
Total land	8735	-
Not degraded	6770	0
Degraded	1965	-
Lightly	650	5
Moderately	904	18
Strongly	411	50

This is understandable from the comparatively small areas occupied by mining and its temporary character of land occupation.

Mining is defined (Down and Stocks, 1978) as 'the removal of materials from the earth's crust in the service of humans'. It is an extractive industry, which, by withdrawing the raw materials from the land, creates pits, and scars leaving the land degraded. Therefore, though its areal extent is often small, its nature is such that land degradation is an inherent part of it.

In India too, mining areas are not commonly included in the discussions of degraded lands. For example, none of the various estimates of degraded land in India (see for an overall view, Chambers *et al.*, 1989) included land that is affected by mining.

Both mining and agriculture are primary activities. As mining yields quicker returns, agriculture always loses in competition wherever extractable resources occur. Therefore, in mining areas, landuse pattern rapidly gives way from agriculture to mining and associated linkages, warehouses, colonies for workers, subsidence-prone areas, dumping grounds for tailings and removed overburden, and if not any one of these, then just derelict land where nothing grows, topsoil is gradually eroded and the land is left barren and fallow (Dhar and Thakur, 1995).

The in-situ materials forming land have been formed over thousands of years through exceedingly slow geological and geomorphological processes. Mining, however, degrades them at an alarming rapidity, making these areas appear like blights on the surface. Post-mining landuse reconstruction through best practices is yet to be introduced in the Indian context.

In this article, we will outline land degradation due to open cast and underground coal mining in the context of a specific region, the Raniganj coalbelt (RCB) of eastern India.

Mining in the Raniganj Coalbelt

The Raniganj coalbelt comprises a total area of nearly 1,260 sq. km. and is located about 250 kilometres north-west of the Calcutta metropolis. The area is bounded by 23°33′ N and 23°53′ N latitudes, and 86°37′ E and 87°23′ E longitudes. A major part of this coalfield falls within Burdwan district, which we have identified as our study area (Figure 1).

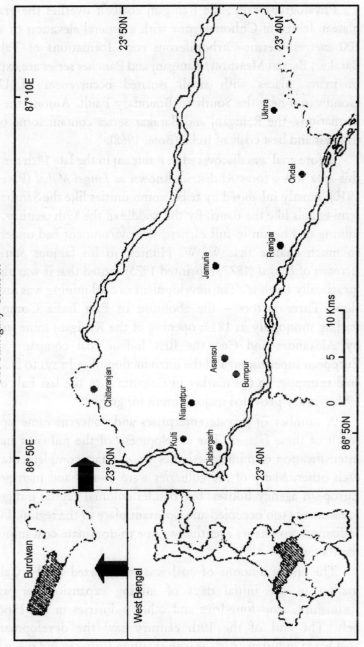

Figure 1
The Raniganj Coalbelt

Physiographically, the Raniganj coalbelt overlies the granitic plateau fringe of Chhotanagpur with a general elevation of about 100 metres. Permo-Carboniferous rock formations of Talcher, Barakar, Barren Measures, Raniganj and Panchet series are exposed in many places with small isolated occurrences of Upper Gondwanas near the Southern Boundary Fault. Among the rock formations, the Raniganj and Barakar series contain some of the thickest and best coals of India (Bose, 1968).

Before coal was discovered in Raniganj in the late 18th century, this area was a forested district known as *Jangal Mahal* (Paterson, 1910), mostly inhabited by tribal communities like the Santhal and semi-tribals like the Bauri. By the middle of the 19th century, after mining had begun in full earnest, its environment had undergone so much change that W. W. Hunter, in his famous *Statistical Account of Bengal* (1877, reprinted 1973), noted that it was already 'practically treeless'. The development of coal mining was initially slow. Three factors -- the abolition of East India Company's trading monopoly in 1813; opening of the Raniganj mine in 1920 by Alexander and Co., the first Indian coal company under European supervision; and the introduction of railways to facilitate coal transport to the market in Calcutta – in the last half of the 19th century provided major stimuli for growth.

A number of private enterprises and concerns came up as a result of these factors. The development of the railways and the intensification of mining activities were in fact complementary to each other. Many of the collieries were owned and managed by European agency houses, but besides colonial capital indigenous investments too occupied an important place in the region. In fact, Indian entrepreneurs eventually came to dominate coal mining in the Raniganj.

The entire amount of coal was transported to the Calcutta market in the initial days of mining expansion for use in steamships, iron foundries and other industries in the Hooghly belt. The end of the 19th century saw the development of coal-based industries such as iron-smelting furnaces and paper mills in the region itself. Still, mining continued to remain an 'enclave' industry (Rothermund and Wadhwa, 1978), serving the primary metropolitan enclaves.

With independence, the pace of mining became faster. To meet the increased demand, private entrepreneurs intensified their mining operations in this region, and a large number of collieries mushroomed in the region in an unplanned way. To prevent this haphazard development and to promote planned growth of coal mining, the collieries were nationalised in several phases between 1972-73 (Basu, 1984). Since then, the Eastern Coalfields Limited (ECL), one of the eleven subsidiaries of Coal India Limited (CIL), has been almost solely responsible for coal mining activities in this area (excepting a couple of BCCL and private collieries). ECL is now the largest landowner and employment provider in the region, and is directly responsible for the impacts of coal mining.

In this paper, we concentrate our discussion on a few issues related to land degradation in the RCB. These are: (i) decline of agriculture, (ii) forestry, and (iii) dereliction of land due to coal mining.

Decline in Potentiality of Land

Before going to discuss the decline of productive capacity of land, we draw a picture about land-man ratio of the RCB when land is treated as 'capital'. As population pressure shows increasing trend, the land-man ratio shows a decreasing trend. Table 2 shows the exponential decrease of land per person in the RCB.

Table 2
Land-man Ratio of the RCB

Year	Land per person (hectares)
1951	0.1905841
1961	0.1229950
1971	0.0970226
1981	0.0749475
1991	0.0576830

The exponential trend curve, is obtained as:

$$LP_t = 0.09967 \; e^{-0.28866t} \; ... \tag{i}$$

with standard error (s.e.) 11.36%. Where LP_t denotes land per person at time 't', and 't' denotes time period with origin at year 1971.

Population pressure of the RCB is so high that the amount of cultivable land per capita will fall rapidly in near future. The projected decline rate in potentially cultivable land of the RCB is very high in comparison to some selected continents. Table 3 gives a glimpse of that alarming feature.

Table 3
Projected Decline in Usable Land Per Person, 1990-2025

Continents/Region	Land per person (hectares)		Decline rate
	1990	2025	%
Central and South America	2.00	1.17	41.50
Sub-Saharan Africa	1.60	0.63	60.62
West Asia and North Africa	0.22	0.16	27.27
South and South-east Asia	0.20	0.12	40.00
The Raniganj Coalbelt (RCB)	0.60[*]	0.02[*]	66.67

Source: Stocking (2000)

[*] Obtained using equation (i)

Urbanisation of the RCB

Lahiri-Dutt (1996, 1994, 1994) and Jana, *et al.,* (1998) described the nature and extent of urbanisation of the RCB from different points of view. They pointed out that the area has a higher level of urbanisation than not only the national average but also those of the state of West Bengal and the district of Burdwan.

In the RCB, 30.87% of the total population lived in 9 urban cities in 1951, but in 1991 about 69.27% of total population lived in 53 cities and towns of various size. This increase in the level of urbanisation has been accompanied by a rapid physical expansion of urban area.

Table 4
Urban Area of the RCB

Year	Area (square km.)	% of total area
1931	24.60	1.95
1941	31.73	2.52
1951	64.40	5.10
1961	169.82	13.84
1971	256.89	20.38
1981	369.29	29.30
1991	561.64	44.58

Table 4 indicates the quadratic trend of urban area growth in RCB given by:

$$A_t = 12.01 + 7.03t + 1.2t^2$$

where A_t = % area of total area at time 't' and t = time period with origin at 1961 (If the expansion rate of urban area remains the same, the RCB will be totally urbanised by the year 2025). The projected urban area will be 106.154% of total area in the year 2025.

Decline of Agricultural Production

The RCB consists mostly of rocky and rolling country scattered with coal pits and factories. However, it is rich with the deposition of the washed up silt from the hills of the Santhal Parganas, Singbhum, Manbhum and Chhotanagpur plateau. The *aman* rice, grows over an area larger than other crops like *aus* and *boro* rice, and pulses. It is generally cultivated on lowlands with a clayey soil, which can accumulate water during rains and remain under water till the end of autumn. Since the *aman* rice is cultivated throughout the region, we mention only the production of *aman*. The district agricultural office provides the following Table 5.

Table 5
Area and Production of Aman of the RCB

Year	Area (thousand hectares)	Production (thousand tonnes)	Yield rate (kilogram per hectare)
1986-87	45.308	131.510	2902
1987-88	32.843	44.440	1353
1988-89	45.984	68.950	1499
1989-90	47.815	74.590	1550
1990-91	49.636	83.480	1680
1991-92	44.876	75.160	1685

Conclusion

In RCB, coal mining over a long period has led to the degradation of land in various ways. The most conspicuous impacts are of course physical, such as the removal of forest cover and soil erosion as well as deterioration in water and air quality. Equally, if not more, significant are changes in the social and economic composition and nature that are inextricably tied to physical changes.

In this region, accelerated mining has led to a high level of urbanisation that entails a wider sprawl of built-up area. At the same time, there has been a decline in land potentiality and agricultural production.

Coal mining by nature is a temporary occupation of land. With increased mechanisation, collieries in Raniganj are becoming more efficient and larger. The region is already dotted with abandoned collieries and pits, and the presence of underground voids of unknown depth often leads to land subsidence and mine fires.

Mining can never be a sustainable activity. However, one can attempt to build up sustainable economies in mining regions through the active involvement of local communities in every aspect of mining. The world mining community continually debates the issues related to degradation, and several guidelines have emerged as best practices in recent years. India can now take notice of these practices and set up its own best practice rules to be followed by both public and private entrepreneurs.

References

Basu, M. (1984) 'The Damodar Valley Coalfields: A Geographical Study' in A. Ramesh (ed.) *Resource Geography*, Heritage, New Delhi.

Bose, S.C. (1968) *Geography of West Bengal, India: The Land and People*, National Book Trust, New Delhi.

Brookfield, H. and Padoch, C. (1994) 'Appreciating agrodiversity: A look at the dynamism and diversity of indigenous farming practices', *Environment*, 36(5), 6-11, 37-45.

Chambers, R. (1986) Sustainable livelihood: An opportunity for the World Commission on Environment and Development, IDS, University of Sussex.

Dhar, B.B. and D.N. Thakur (1995)(eds.) *Mining Environment: Proceedings of the First World Mining Environment Congress*, Oxford and IBH Publishing Co. Pvt. Ltd., Calcutta.

Down, C.G. and Stocks, J. (1978) *Environmental Impacts of Mining*, Applied Science Publishers Ltd., London.

Enters, T. (1997) *Methods for the economic assessment of the on and off-site impacts of soil erosion*, Issues in Sustainable Land Management, No. 2., International Board for Soil Research and Management.

Hunter, W.W. (1877, Reprinted 1973) *A Statistical Account of Bengal*, Vol. IV, D.K. Publishing House, Delhi.

Jana, P.K., M. Mukherjee and K. Lahiri-Dutt (1998) 'Nature and extent of urban sprawl in the Raniganj coalbelt, West Bengal: Use of satellite images and census figures as databases', *Indian Cartographer*, Vol. 18, 201-207.

Lahiri-Dutt, K. (1994) 'Mining development and consequent degeneration of the urban environment: A planned alternative for Raniganj town', *Geographical Memoirs*, NBU Journal of Geography, Shillong.

Lahiri-Dutt, K. (1994) 'Urban growth in West Bengal: A study of outgrowths, 1981 and 1991', *Urban India*, NIUA, Delhi.

Lahiri-Dutt, K. (1996), 'Mining Development and Gender in Raniganj Coalbelt', *Transactions*, IIG, Pune, 18(1), 7-15.

Oldeman, L.R., R.T.A. Hakkeling and W.G. Sombroek (1990), *World map of the status of human-induced soil degradation: An explanatory note and 3 maps*, International Soil Reference and Information Centre, Wageningen.

Paterson, J.C.K. (1910), *Bengal District Gazetteers: Burdwan*, Bengal Secretariat Book Depot, Calcutta.

Rothermund, D. and D.C. Wadhwa (eds.) (1978), *Zamindars, mines and peasants: Studies in the history of an Indian coalfield and its rural hinterland*, Manohar Publications, New Delhi.

Stocking, M. (2000, second edition) Soil erosion and land degradation, in Timothy O'Riordan (ed.) *Environmental Science for Environmental Management*, Prentice Hall, London.

11

Mining of Coal and Land Degradation: A Review

Debasish Sarkar

A study by the representatives of the World Bank, observed that "land degradation, whether caused by water erosion, wind erosion, salinisation, waterlogging, nutrient loss, compaction, or overgrazing is extensive in India. Overall, about 163 million hectares of land have some degree of degradation, which represents 49.7% of India's total area. One survey of just agricultural land indicates that 83.4 million hectares of agricultural land are degraded, with 35% being slightly degraded, 31% moderately degraded, and 34% severely degraded. The estimated productivity losses due to land degradation fall between 4% and 6.3% of total agricultural output per year, which represents annual foregone production of $1.5 to $2.4 billion". Keeping in mind the gravity of the crisis, here an attempt will be made to review the scenario of land degradation relating to coal mining activity in India.

Exploration and exploitation of minerals is very essential for the economic and industrial development of the developed and developing countries. Mining activities are generally accompanied by a variety of environmental problems. Mining industry is next only to agriculture and is spread over vast areas throughout the world. The area disturbed by mining is approximately 924,000 hectare per year and therefore at the end of this century, mining alone will degrade nearly 24 million hectare or 0.2% of the earth's surface (Soni and Vasistha, 1987). In India, an area of 683,671.5 hectare is under mining leases spread over 19 states. Mining operations ravage large areas and leave them waste. The mining activities for more than 200 hundred years, involving extraction of material from surface and underground, their processing, transportation and accumulation, have produced notable changes on the surface features, by scarification, creation of spoil tips and subsidence zones, all of which constitute the anthropogenic features superposed on the natural landscape. These features will be discussed with the broad understanding of the concept of 'environmental impacts'.

Land Degradation

The land as an essential resource has its particular importance in an overpopulated country like India where most of the people depend on agricultural activities. Land is a non-renewable asset and its degradation, therefore, has far-reaching implications affecting the lives of thousands of inhabitants living in the degraded land. The critical problem of mining is that it essentially involves land degradation. Land loss and land degradation due to mining are related with the nature of the technology of extraction of coal. Coal mining operations can be considered under two broad categories - (i) *Underground or Deep Mining*, where the coal is extracted from the seam without removal of overlying strata. In this process subsidence causes the major loss of land. The whole area under the operation may be considered to be degraded due to instability created with the underground mine, through the extraction of coal by depillaring and consequent formation of goaf areas. Mine fire is another prime source of degradation of land. Smoke and gases emanate through the cracks or abandoned underground mine openings indicating old mine fires in underground goafs. The land in and around areas looses the agricultural activities. (ii) *Opencast Mining*, where the overburden are removed and the coal is extracted from the exposed seam. The excavation and spoil heaps is the major cause of land degradation. The German scholar Georgious Agricola had put in his 1550 treatise on mining: "The fields are devastated by mining operations ...the woods and groves are cut down...for there is need of endless amount of wood and timbers,... And when the woods and groves are felled, then are exterminated the beasts and birds... Further, when the ores are washed the water which has been used poisons the brooks and streams and either destroys the fish or drives them away." From the statement one can understand the diverse implications of the opencast mining. With increased mechanisation, the effect of mining has increased manifold. Large-scale land sacrifice with the development of spoil dumps are the most significant impacts of opencast mining in this coalfield; it changes the topography of the adjacent land area. Land available for agricultural activity also decreases. The land loss and land degradation due to the opencast mining is much higher than the underground mining.

Raniganj Coalfields: A Case Study

The coal mining in India was first started in the Raniganj Coalfields of the then Bengal. The first Englishman to discover the existence of coal in Bengal was probably Mr. Suetonius Grant Heatly who, in 1774, was the Collector of Chota Nagpur and Palamau. In that year he with Mr. John Summer obtained a license from Warren Hastings, empowering them to work coal mines in Pachete and Birbhum.

Mr. Redferne subsequently joined the firm which as Summer, Heatly and Redferne applied for and obtained the exclusive right, for a period of 18 years, to work and sell coal in Bengal and its dependencies (Bengal District Gazetteers, Burdwan by J.C.K. Peterson). Since then a large number of mines including both opencast quarries and underground mines have been in operation, producing several million tonnes of coal every year. The Raniganj coalfields, the largest single belt of mining operation occupy an area of about 1,000 square kilometres (classically, however, it includes 1,530 square kilometres area in four districts of West Bengal, viz., Barddhaman, Birbhum, Banker, Purulia and two districts of Bihar). The long continued mining operations have resulted in significant changes in the natural environment. It is well known that both underground and opencast mining operations are actively pursued in the recently cleared forest area; while the former invariably leads to gradual subsidence; loss of soil nutrient, and quick run-off of rainwater gradually turn the overground land surface into unculturable fallows; the opencast mines on the other hand, are apt generate voids and overburden dumps. Both the processes affect the land resources and the pattern of land utilisation. An overall impact assessment of degradation of land was made by The Geological Survey of India in 1986 (Records Vol. 114, pp.19-34). According to this report, "the extraction of coal both by opencast and underground methods over the last two hundred years has resulted in large scale defacing of the natural landscape in many parts of the Raniganj Coalfield. Different types of anthropogenic landforms including mine pits, spoil tips and subsided lands have developed on a flat to gently rolling pediplain area". The study noted the following observations relating to the land degradation:

(a) Opencast mining: (1) From the old abandoned quarries and associated spoil dumps, the area occupied is about 16.25 square kilometre. (2) Recently abandoned quarries and associated spoil dumps which have stopped operation about one to three years ago have occupied around 3.75 square kilometre. So, total area occupied is approximately 20 square kilometre. (3) From active quarries and the associated spoil dumps the occupied area is of about 10.5 square kilometre.

(b) Underground mining: Land subsidence damage is the major associated hazard relating to the underground mining. According to the estimates, a total area of approximately 25 square kilometres has been damaged due to land subsidence in this coalfield. But, the master plan of CMDPI/ECL for Eastern Coalfield Limited (ECL) indicates the size of land area is approximately 42 square kilometre.

Estimation of Land Degradation

If we note the proportion of coal production following the basic two methods of extraction then one can note that in the pre-nationalisation and immediately after nationalisation period, ratio was high in favour of underground mining where the damage cost due to degradation of land is much less than that of opencast mining. In 1975-76, at the all-India level, the percentage of underground production was approximately 73.28% where as in 1997-98 the ratio reduced to only 20.76%. In case of Raniganj Coalfield which is mainly under the ECL, the proportion of underground mining was 89.99% in 1975-76 and this ratio decreased to 46.1% in 1997-98. Since opencast mining is basically mechanised in its character therefore the output per manshift (OMS) is high enough to substantiate the rising trend of opencast production. Following the past trend of land degradation, in case of opencast method the estimated size of degradable land increases to 81 square kilometre whereas for underground, the calculated area of land subsidence is reduced to 19.28 square kilometre. Combining these two, the area of degradable land increases manifold. If the same proportion of loss of land is maintained at the all-India level, then in 1997-98 the total area of degradable land will be estimated to around 83 square kilometre for underground mining and approximately 1131 square kilometre in case of opencast mining. So, as the mining of coal increases the size of deprivation of land from agricultural activity increases, also raising the question of sustainable development of the economy. There is a clear trade-off between production cost and environmental cost. So a master plan is to be chalked out in such a way that requirement of coal can be assured with the maintenance of healthy environmental ambience.

Policy Prescription

To maintain sustainable development, extraction of coal and measures to control land degradation should move hand in hand. Land reclamation in a mining area is a vital aspect of environmental management where the basic objective is to eliminate the damages done on the land during mining and prepare the land as per the need of the society. Nature of land degradation varies from opencast to underground and as a result the approach to counter the problem will differ from one method to another. Subsidence of land is common with the underground technique, where the suggestion is to systematically fill the underground goaf with a mixture of sand and water. In case of opencast mining a systematic coordination between pre-operational, operational and post-operational phases of mining is essential. The planning for such a programme should include the demarcation of areas for stock piling of topsoil, sub-soil and overburden separately so that during the operational stage the reclamation work by backfilling of overburden/waste material and the restoration of the topsoil can be taken up simultaneously with the mining.

At the end, one can conclude that though coal is an essential input in the process of economic development but the extraction of coal is creating a series of environmental hazards including degradation of the existing land and therefore, *a micro level planning and its execution* for restricting land degradation at each of the mine is essential to maintain an optimal rate of extraction of coal.

References

Brandon, C. and Hommann, Carter (1995), The cost of inaction: Valuing the economy-wide cost of environmental degradation in India, presented at the "Modeling Global Sustainability" conference, United Nations University, Tokyo.

Chowdhury, S. Roy and Roy, U.K., Geological Survey of India, Records Vol.114.

Peterson, J.C.K. (1997), Bengal District Gazetteers, Burdwan.

12

Land Degradation and Human Action: A Case Study in Burdwan

N.K. De and Ananya Taraphder

Land degradation is the result of different processes acting upon the uppermost layer of the earth surface. A piece of land surface is the end product of geology, parent material, geomorphology, climate, soil and land forming processes, operating through time. Man and the natural processes acting on land surface are responsible for land degradation. Land degradation at the present juncture, has become an important aspect of study of all scientists in general. Land degradation is taking place all over the earth surface.

About 40% of the world's population is inhabiting in the tropical areas (30°-35° North and South) and have agro-based economy. Too much population and negative man-land ratio in the tropics create pressure on land, where both natural and human activities are responsible for land degradation.

In the present paper a case study has been made in Burdwan— a tropical environment, where almost every year recurrent flood and drought are common. In the district, monsoonal heavy downpour in short duration is responsible for recurrent floods, change of river courses, channel geometry and dynamics. In addition, human actions are further aggravating the situation.

Study Area

The district of Burdwan lies between 22°56'N and 23°53'N latitudes and 86°48'E and 88°25'E longitude. The district has an area of 7028 sq. km. Physically it can broadly be divided into three parts. The western part having higher relief shows a barren look. The eastern part with highly fertile soil presents almost a level surface. In between lies the rolling upland, which gradually merges with the monotonously flat plain in the east. The economy of the district depends both on agriculture and industry. The western half is dominated by mining and industrial activities, whereas the eastern half is mainly agricultural (Figure 1).

Methodology

The methodology followed in the present paper may be divided into three: Pre-field method—consists of collection of secondary data and information from journals, reports, available literatures (both published and unpublished); Field method—includes collection of primary data from some selected areas, which have been subsequently verified also with the secondary information, already collected; Post-field method—includes arranging, processing, tabulating and analysis of the data, presenting them in pictorical forms and preparation of final report.

Factors of Environment and Land Degradation

Land degradation in the district of Burdwan is due to the actions of running water (surface runoff and channelised water) and by human activities. As most of the rivers of Burdwan are conditioned, human actions are dominating the area. As such both physical and cultural factors are influencing the water management.

As stated earlier, the district of Burdwan can be divided into three regions—the western upland terrain/plateau fringe area, the central rolling plain and the eastern flat plain. Relief is maximum in the west. The central rolling plain has an uneven character with less variation in relief. The eastern monotonous flood plain is drained by major rivers—Bhagirathi, Ajay and Damodar. The

Figure 1
Location Map

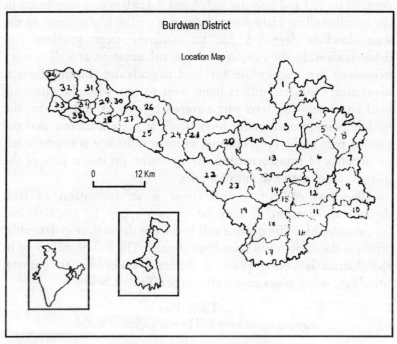

1. Ketugram-I	11. Memari-I	21. Ausgram II	31. Barabani
2. Ketugram-II	12. Memari-II	22. Galsi-I	32. Salanpur
3. Mangalkot	13. Bhatar	23. Galsi-II	33. Kulti
4. Katwa-I	14. Burdwan-I	24. Kanksa	34. Asansol
5. Katwa-II	15. Burdwan-II	25. Durgapur	35. Hirapur
6. Manteswar	16. Jamalpur	26. Faridpur	36. Ghittaranjan
7. Purbasthali-I	17. Raina-I	27. Andal	
8. Purbasthali-II	18. Raina-II	28. Raniganj	
9. Kalna-I	19. Khandoghosh	29. Jamuria-I	
10. Kalna-II	20. Ausgram-I	30. Jamuria-II	

general slope is from west to east. Climate is in general hot-humid (monsoon) type. Rainfall is high in the east (152 cm), which decreases towards west (132 cm). West has a stony and skeletal soil. Soil is clay loam to sandy loam in the east. In between, the area is covered by red and lateritic soil. Land degradation is maximum in the middle rolling plain due to human activities like mining. In the west, land is degraded due to relatively steep gradient and deforestation. In the east, agriculture, urbanisation as well as river behaviour are responsible for land degradation. In the district, vegetation character differs from west to east with the changing land character----western part covered by grasses and scrubs, the middle portion once under forest (Sal), at present barren, and the eastern part under crop culture. Human influence is strongly felt on account of transformation of vegetation (in major part of the district) from native to cultural ones.

The loss of surface soil cover is an indication of land degradation/erosion. It may be of two types: (a) normal, and (b) catastrophic. The normal soil loss from the surface is desirable, whereas the catastrophic one is unwanted. The loss of soil cover in the district is taking place at different scales due to climate, lithology, relief, plant cover, etc., as per Table 1 below.

Table 1
Factors Influencing Soil Loss at Different Scales

Scale of Analysis			Evidence
Macro	Meso	Micro	
Climate	Lithology, relief		Sediment yield of rivers
Climate	Lithology, relief	Micro climate, lithology, soil	Drainage density
Climate	Altitude, relief		Study of erosion rates
Climate		Plant cover, micro climate	Studies of soil loss from hill-slopes

Based on R.P.C. Morgan (1986) "Soil erosion and conservation".

Land Degradation by Man-Induced Water Action in Burdwan

Burdwan is blessed by a number of rivers like Damodar, Bhagirathi, Ajay, Bhalluka, Khari, Behula, Barakar, Kunur, Kana,

Ghea, etc. Agriculture has flourished due to favourable agro-climate conditions (soil, climate, etc.). But, Damodar with its devastating floods almost in every alternate year has a negative impact on the economy. Even before and just after independence agriculture in the district was mainly dependant on natural sources of water and accordingly agriculture was basically rainfed. The availability of water in a river (river discharge) is the result of rainfall in the catchment area as well as the rate of evaporation (depending on temperature).

The amount of rainfall in the catchment areas of the Damodar system and the corresponding amount of runoff is explained in Table 2.

Table 2
Annual Runoff (1901-1950) of the Damodar River System

Name of Rivers	Catchment Area (sq. km.)	Normal Annual Rainfall (mm)	Mean Temperature (in °C)	Annual Runoff (million cubic m)
Damodar	16,509.6	1,313.0	26.2	8,103.3
Ajay	6,036.0	1,268.3	26.8	2,639.6
Khari	2,353.6	1,352.8	27.2	1,036.1
Hoogly and Others	3,179.9	1,549.9	27.0	1,739.2

Source: Damodar Valley Planning Atlas (1969).

Table 3
Average Surface Runoff in the District of Burdwan (1901-1950)

(in cm)

Area	Place Names	June	July	August	September
Southern Burdwan	Burdwan	10.74	19.15	17.75	7.42
Eastern Burdwan	Kalna	8.56	14.33	11.53	5.33
Northern Burdwan	Monteswar	9.35	14.30	17.75	4.37
North-East Burdwan	Katwa	9.88	14.30	13.87	4.52
Western Burdwan	Asansol	9.27	22.05	21.01	7.47

Source: Damodar Valley Planning Atlas (1969).

Further rainfall affects soil erosion and hence land degradation. Sometimes soil moisture deficiency prevents runoff. Again runoff in the area depends on the nature of parent material, permeability and porosity of the rocks. Table 3 explains the average runoff in different parts of Burdwan.

The way soil moisture deficiency arrests runoff in Burdwan can be explained by Table 4 below.

Table 4
Soil Moisture Deficiency in Burdwan in Different Months

Station	January	February	March	April	May	October	Nov.	Dec.
Burdwan	-8.15	-7.11	-11.56	-10.24	-0.89	-5.33	-9.98	-9.09
Asansol	-7.29	-7.19	-11.63	-12.83	-8.79	-1.88	-9.47	-8.53

Source: Damodar Valley Planning Atlas (1969).

Normal seasonal floods followed by droughts were prevailing in Burdwan till 1946. Damodar Valley Corporation (DVC) was the first human imprint to regulate the natural flow of the river. With the advent of green revolution, transformation of agriculture (due to population pressure) took place not only in the use of water by taking recourse to canal irrigation by constructing dams, but also in other aspects of advanced agricultural technology (since 1966-67).

The devastating floods of 1978 and 2000 in Burdwan and the adjoining areas (Damodar Valley region) were the result of deforestation (in the catchment area of Damodar), construction of big dams (Maithon, Tilaiya, Konar, Panchet----Figure 2), barrage (Figure 2), reservoirs and embankments in the DVC region. This has resulted in the decrease in sediments in the downstream section of the channels and valleys of the Damodar system. In addition, coal mining activities in the west have further aggravated the situation.

The severe cyclonic storm in 1978 resulted in heavy downpour of 600 mm in Burdwan between September 26-29. In the year 2000, intense depression gave rise to about 1200 mm rainfall in Burdwan and surrounding region which is much more than the rainfall in West Bengal since 1950's.

The sedimentation rates in different dams of the Damodar Valley Region (DVR) have superceded the expected rates in the current years due to faulty management and planning (Table 5). Increased siltation of the river channels results in flood (Figure 3).

Figure 2
Damodar Valley Project

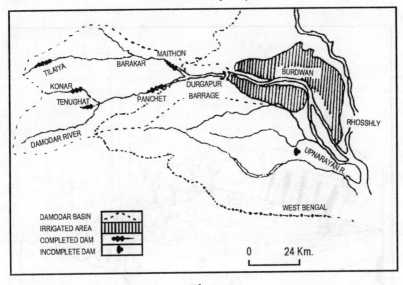

Plate
*Durgapur Barrage Facing Downstream of Damodar River
(Embankments are seen on both banks)*

Figure 3
A Block Diagram Showing Siltation in River Channel and Resulting Flood

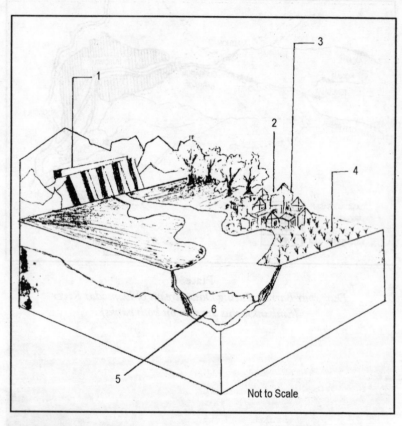

1. Increasing rate of sedimentation after construction of dam & reservoir.
2. Settlements.
3. Excessive rain creating flood.
4. Cultivated land.
5. Sedimentation and lack of dredging decreases depth of channels.
6. Silt.

Table 5

Expected Rate of Sedimentation in Reservoirs of DVR

Name of Dam	Prospected Sedimentation (acre foot/year)	Actual Sedimentation (acre foot/year)
Maithon	684	5980
Panchet	1982	9553
Mayurakshi	538	2000

Land Degradation by Other Human Actions

Land is degraded not only by water action (both natural and human) but also by other human activities in Burdwan, such as:

1. Mining activities
2. Industrial activities
3. Agricultural activities
4. Municipal activities

Strategies to Combat Problems Resulting from Human Interferences

Strategies proposed to overcome land degradation by different human activities in Burdwan are as follows (refer Table 6):

1. Mining Activities:
   a. Construction of concrete pillars and sand filling in abandoned mines to check land subsidence.
   b. Void pits filled up with water can be used as fishing grounds.
   c. Water treatment plant in coal mines is necessary to check deposition of coarse sediments in channels/valley bottoms of Damodar river.
   d. Sand stowing necessary in coal mines.
   e. Arrangement for regular dredging of river channels, etc.
2. Industrial Activities:
   a. Fly-ash can be used to make bricks in Durgapur area. It can also be used as a material for doing concrete.
   b. Increase in height of chimneys and use of filters.

Table 6

Man-Induced/Human Activities and Resulting Problems in Burdwan

Types of Activity and Area	Nature of Activity	Effects	Problems/Pollutants
Mining (Raniganj, Jamuria)	Directly related to land (sub-surface); Opencast mining—rare, pit mining—in general; River water used for cleaning (coal washery)	Creation of hollows (underground); Inversion of topography and material; Deposition of large quantities of coarse sediments in channels/valley bottoms	Land subsidence; Loss of agricultural land; Water Pollution from coal washery; Recurrent flood in D.V.R.
Industrial (Durgapur, Asansol, Raniganj, Hirapur, Jamuria)	Related to land, air and water; Iron & steel and agrobased industries established	Emission of CO_2 CO, SO_2; Industrial waste water with ions of chlorine, sulphate, bicarbonate, nitrate, potassium and heavy metals; Fly-ash and other soild wastes of D.P.L. and D.S.P.	Air pollution; Contamination of river water and as a result loss of top soil; Loss of soil fertility in agricultural fields.
Agricultural (Most of the blocks in central and eastern Burdwan)	Totally land based (water related) activity. Mostly multiple cropping—rice, potato, mustard, wheat, poppy seeds, etc.	Introduction of packing practice—high requirement of chemical fertilizer, pesticide, water hence construction of dams, barrages and reservoir; Draining of excess water from agricultural field into rivers (Khari river basin) and underground (by digging ditches etc.)	Accumulation of toxic chemicals on surface, change in physical and chemical properties of soil; Contamination of flowing & ground water; Increasing siltation and disturbing environments balance by creating hydrostatic pressure on underlying rocks through dam construction; Lowering of water table etc.
Municipal (in urban areas)	Related to land, air and water, water supply, waste disposal—main activity	Ground water tapping & supply through plastic pipes; Dumping of solid wastes (plastic, paper, metal cans, glass bottles, aluminium foil, metal junks, domestic garbage) in same area	Subsurface—exhausted water layers & resulting subsidence; Carcinogenic effect of plastic water supply pipes; Slow process of humification; Shortage of drinking water etc.

    c.   Fly-ash can be used in the cultivation of wheat, potato and poppy seeds, etc. as fertilizer.

3. Agricultural Activities:

    a.   Encouraging farmers to use green manures instead of chemical fertilizers.

    b.   Introduction of pest-resistant crop to get rid of chemical pesticides necessary.

    c.   Overcoming land salinisation due to canal irrigation, contour bunding and construction of small dams and storage of surface water as well as restoration of ground water potential needed.

    d.   Use of biofertilizers for sustainable agricultural development.

4. Municipal Activities:

    a.   Supply of drinking water with the help of water treatment plant by using river water (Banka, Damodar, etc.)

    b.   Imposing ban on the use of plastic/polythene.

    c.   Different types of garbages can be dumped in different areas for recycling.

These are the few major strategies which should be taken care of immediately to fight against land degradation by human actions.

Concluding Remarks

From the above discussion it is evident that the human activities are much more responsible as compared to natural ones, in the district of Burdwan, for the existing land degradation problems. It is, therefore, necessary to control human actions not only by enacting laws but also by involving local people by creating awareness among them. The government concerned should be sincere in these aspects, otherwise, the land degradation problems are bound to aggravate further in future.

References

Arvill, R. (1967) *Man and environment: Crisis and the Strategy of Choice*, Penguin, Harmondsworth.

Bryan, R.B. (1981) Soil erosion and conservation, in *Man and Environmental Processes*, edited by K.J. Gregory and D.E. Walling, Butterworths, pp. 207-221.

Crossland, J. (1978) Reporting pollution, *Environment*, Vol. 20, pp. 29-31.

Dassman, R.F. (1975) *The Conservation Alternative*, Wiley, New York.

Detwyler, T.R. (1971) *Man's Impact on Environment*, McGraw Hill, New York.

Dobzhan Sky, T. (1950) Evolution in the tropics, *American Scientist*, Vol. 38, pp. 209-221.

Gregory, K.J. and Walling, D.E. (1981) *Man and Environmental Processes*, Butterworths, London.

Gupta, H.K. (1976) *Dams and Earthquakes*, Elsevier, Amsterdam.

Lee, N. and Wood, C. (1972) Planning and pollution, *The Planner*, Vol. 58, pp. 153-158.

Moore, N.W. (1968) Experience with pesticides and the theory of conservation, *Biological Conservation*, Vol. 1, pp. 201-207.

Morgan, R.P.C. (1986), Soil erosion and conservation, edited by D.A. Davidson, Longman - Scientific and Technical, London.

Park, C.C. (1980) *Ecology and Environmental Management*, Butterworths, London.

Singh, S. (1997) *Environmental Geography*, Prayag Pustak Bhawan, Allahabad.

Technical Advisory Committee (1969) The Planning Atlas of the Damodar Valley Region.

13

Relationship Between Slope and Agricultural Landuse in the Damodar-Dhajmma Interfluve of Burdwan and Purulia Districts

Nageshwar Prasad and Manjari Sarkar (Basu)

Slope is one of the most important aspects of geomorphology. Slope is the main property of landforms that influence agricultural activities, different landuse, transport and settlement also. Available resource of any area depends mainly upon the nature of landscape. Land capability is generally assessed by evaluating each parameter of terrain. The physical parameters which are taken into account are surface elevation, slope, ruggedness index, drainage matter, geology, etc. If a single factor amongst these is to be considered as the basic determinant of landuse, it will most certainly be slope. This paper has, therefore, made an attempt to study the role of slope in agricultural landuse in the Damodar-Dhajmma interfluves.

Study Area

The area, Damodar-Dhajmma interfluve, is a quadrant comprising about 78.62 sq. km. It is a transitional zone between the Chhotanagpur plateau and alluvial tract of the lower Ganga plain. River Damodar passes through this area and this river valley is the southern-most boundary of the Burdwan dist. Rest of the areas of this study area fall under Purulia district (Figure 1). This area extends from 23°40'N to 23°45'N and 86°50' E to 86°55' E. It has a contrast between agricultural and industrial land. This area comprises some parts of five police stations, viz., Kulti, Hirapur, Asansol, Salanpur in Burdwan district and Neturia in Purulia district. Total population is about 44,470 (1991 census).

Damodar is the main water resource of this region. It flows over this area from south-west to south and south-east direction. Rivers are mainly controlled by the original slope. Besides Damodar, other small streams are flowing over this region, viz., Dhajmma, Harial, Jhor. Due to greater erosional activities in the upper course, this river exhibits depositional features in the form of bars and deposits of stony waste along the river valley.

The major portion of studied area, mainly in north-western portion which comprises the coalfield area, primarily consists of barren rocky and rolling undulating ground with a laterite soil. The average elevation of this area is 60.96 metres. The highest elevation found in the north-western part of this area is about 152.4 metres; in south-eastern part of this region higher ground is found. Along the Damodar river and south-eastern corner is relatively lower in height. These areas lie between 106.68-91.44 metres elevation. The major slope of this region is in north and north-west direction along the main river (Damodar). Local slope depends upon the tributaries of Damodar mainly in north-south direction. The topography is not natural, it is modified by the human activities mainly in mining and industries. Most of this region is covered by gentle slope ranging from 1.5° to 5°. In the north-western corner slope is steep to moderate, it ranges between 2°-2.5° and above. The average slope varies between 0.5°-2.5° over the whole region. Moreover, the region is lowlying undulating erosional surface.

Climatically this region falls under the tropical climatic condition. But this area has more continental characteristics. The average maximum temperature is higher (39.1°C) and winter temperature is lower (17°C). The average maximum temperature

Figure 1

ranges from 17° to 18°C during winter (November-January) and 34°-39°C during the summer (April-June). There is much less frequency of nor'westers and both annual and summer rainfall is markably less. Relative humidity of this region is approximately 62%. The average annual rainfall is about 1200 mm and major part of the precipitation takes place during the period of June to September. In winter season the amount of rainfall is very low and uncertain. In months of April and May the area is subjected to *Loo* during daytime.

Typical tropical and continental type of climate indicates that the natural vegetation of this area is mostly tropical thorny type. They consist of open stunted and widely scattered trees and bushes. Date palm *(Phoenix sylvestris)*, acacias *(Acacia arabica)*, neem *(Melia azedarach)*, khair *(Acacia calechu)* are the major species. Apart from these, open Sal *(Shorea robusta)*, Mohuya *(Bassia latifolia)*, Palas *(Butea frondosa)* forest thrive mainly on laterite soil, and dense Sal forests on red and yellow loams in the north-western part of this region. These forests have economic value but most of the forests have been cleared for the industrial development and urban agglomeration.

Soils of the study area vary with degree of slope and geomorphological locations. Mainly three types of soil are found in this area, namely, such as laterite soil, red and gravelly soil, and alluvial soil. Alluvial soil is found along the drainage line. Fine texture soil is found along stream courses, but away from drainage line the texture becomes coarser and sandy loams are prominent. Laterite soils are found in the north and north-western part of the study area; these soils are infertile. Red and gravelly soils are found in the southwestern corner of the study area. These soils have a coarse texture and are primarily sandy. For crop production in these soils enough water supply must be available. The fertility of this soil can be raised by the application of fertilizer and sufficient irrigation.

Methodology

For establishing a relationship between slope and agricultural landuse, two sets of parameters were firstly prepared. At first the whole area was divided into slope categories (Figure 2) following Raisz and Henry method. Secondly, on the basis of census reports (village-wise) two main landuse categories—irrigated area and non-irrigated area (Table 1)—were computed. Since the agricultural data were computed village-wise, average slope of each village was needed. To obtain this average value of slope, the village map was superimposed on slope category map. Obviously, most of the slope categories have more than one type of agricultural areas. So the

Figure 2

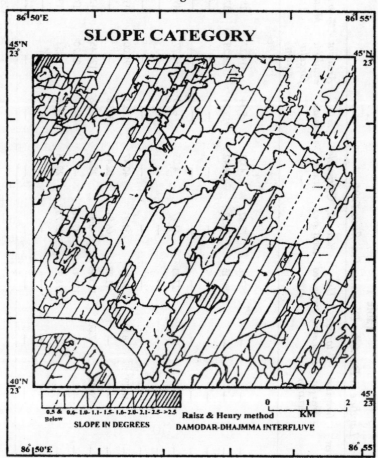

SLOPE CATEGORY

86 50'E — 86 55'

23 45'N — 23 45'N

23 40'N — 23 45'

| 0.5 & Below | 0.6- 1.0 | 1.1- 1.5 | 1.6- 2.0 | 2.1- 2.5 | >2.5 |

SLOPE IN DEGREES

Raisz & Henry method

0 1 2 KM

DAMODAR-DHAJMMA INTERFLUVE

86 50'E — 86 55

authors made the averages of cultivated areas (units) in each slope category for better correlation analysis, keeping maximum accuracy in mind.

Slope Categories

Six major slope categories have been found in the study area (Figure 2). The slope map reveals that the major portion of this region is covered by a gentle slope ranging from 0.5° to 2.5°. In the

Table 1
Landuse in Damodar-Dhajmma Interfluve 1991

Name of the villages	Area of mauza (in hectares)	Forest	Irrigated area (in hectares)	% to the total area	Un-irrigated area	% to the total area	Cultivable waste	% to mauza	Area not available for cultivation	% to the Mauza
P.S. Kulti										
Sabanpur	75.78	—	—	—	57.46	75.82	4.05	5.34	14.27	18.83
Barira	201.85	—	—	—	99.15	49.12	36.42	18.04	66.28	32.83
Lalbazar	—	—	—	—	—	—	—	—	—	—
Ramnagar	—	—	—	—	—	—	—	—	—	—
Bali Tara	61.93	—	—	—	27.21	43.77	4.05	6.53	30.77	49.68
Petana	—	—	—	—	—	—	—	—	—	—
Kulti	—	—	—	—	—	—	—	—	—	—
Lachhmanpur	237.62	—	16.19	6.81	82.96	34.91	40.47	17.03	98.00	41.24
Rampur	61.62	—	—	—	41.28	66.91	20.41	33.08	—	—
Chalbalpur	286.36	—	20.23	7.06	125.45	43.80	20.23	7.06	120.45	42.06
Dedi	135.11	—	—	—	121.41	89.86	—	—	23.70	17.54
Kultara	123.72	—	—	—	78.91	63.78	20.23	16.35	24.58	19.86
Punauri	—	—	—	—	—	—	—	—	—	—
Niamatpur	—	—	—	—	—	—	—	—	—	—

Contd...

Contd...

Badrichak	129.94	—	—	—	105.62	81.28	24.32	18.71	—	—
Mahuldi	126.42	—	—	—	59.08	46.73	12.14	9.6	—	—
Shipur	63.36	—	16.19	23.34	49.68	71.62	—	—	11.49	3.41
Gangutia	336.76	—	40.46	12.04	167.94	49.98	40.47	12.04	4.02	6.19
Raydi	64.94	—	20.23	31.15	40.47	62.31	—	—	4.02	23.05
Mahatadi	173.08	—	—	—	133.18	76.94	—	—	4.24	31.16
Jasaidi	126.44	—	8.09	6.39	52.6	41.60	—	—	39.90	31.90
Dishergarh	—	—	—	—	—	—	—	—	65.74	—
Shitalpur	—	—	—	—	—	—	—	—	—	—
Monoharchak	41.15	—	—	—	22.26	50.09	—	—	18.89	16.24
Chhota	112.79	—	—	—	66.77	59.19	—	—	—	—
Dhemua	—	—	—	—	—	—	—	—	—	—
Sodpur	165.67	—	—	—	87.01	52.52	—	—	78.66	46.46
Radhanagar	180.67	—	—	—	87.41	43.35	—	—	93.34	51.64
Asanbani	—	—	—	—	—	—	—	—	—	—
Sitarampur	47.71	—	—	—	22.66	49.57	—	—	23.05	50.42
Belrui	—	—	—	—	—	—	—	—	—	—
Lachhipur	119.02	—	—	—	51.80	43.52	—	—	67.22	56.46
Bamandiha	126.19	—	—	—	54.63	43.29	—	—	71.56	56.70
Aldihi	78.21	—	—	—	30.76	39.33	—	—	47.45	60.66
Methani	205.61	—	—	—	84.17	40.93	—	—	121.44	59.06

Contd...

Contd...

Kamarpur	101.22	—	—	—	45.32	44.77	—	—	55.90	55.22
Herrelgaria	110.69	—	—	—	50.99	46.06	—	—	59.75	53.93
Bajdihi	72.47	—	—	—	27.11	37.40	—	—	—	62.59
Paidi	—	—	—	—	—	—	—	—	45.00	—
Chinakuri	56.77	—	—	—	39.66	69.86	—	—	17.11	30.13
Total	3629.10	—	121.39	0.33	1912.85	55.70	222.79	7.13	1202.71	36.14
P.S. Hirapur										
Jandiha	91.46	—	6.07	6.63	22.66	24.77	50.90	55.31	12.14	13.27
Junut	131.52	—	14.16	10.76	38.45	29.23	36.42	27.69	42.49	32.30
Bhaladi	75.68	—	8.09	10.68	6.48	8.56	25.09	33.14	36.02	47.59
Mamabara	78.10	—	3.24	4.14	70.82	90.67	4.04	5.17	5.17	—
Chapadi	329.01	—	24.28	7.37	189.80	57.68	30.76	9.34	84.17	25.58
Aluthiya	71.23	—	20.28	28.47	10.12	14.20	8.09	11.35	32.79	46.03
Bharatchak	77.70	—	10.12	13.02	7.69	9.89	4.45	5.72	55.44	71.35
Patmohana	204.77	—	16.19	7.90	74.87	36.56	30.35	14.82	83.46	40.70
Bidyanandapur	112.90	—	36.42	28.03	16.19	12.46	20.23	15.57	57.06	43.92
Bara-Dighi	210.84	—	6.07	2.87	28.33	13.43	111.29	52.78	65.15	30.09
Shanmara	76.49	—	22.25	29.08	39.25	51.31	3.24	4.23	11.75	15.36
Purushottampur	76.49	—	30.35	13.08	110.48	47.64	1.21	0.52	89.35	38.74
Lkrasata	231.89	—	12.14	8.30	8.09	5.53	26.30	18.00	99.56	68.14
Total	1785.08	209.66	11.74	623.23	33.91	351.75	19.70	626.84	18.00	34.71

Contd...

Contd...

P.S. Asansol									
Sudi	192.49	—	119.79	62.23	—	12.14	6.30	60.56	31.46
Raghunathbati	85.54	—	—	14.16	16.55	6.07	7.09	65.31	76.35
Ranjibanpur	90.22	3.64	4.03	18.21	20.18	12.14	13.45	56.23	62.32
Baradhemo	339.30	114.12	33.65	—	—	22.26	6.56	202.75	59.78
Jagadi	48.80	1.62	3.31	16.18	33.17	6.07	12.43	24.92	51.06
Bartari	171.07	90.25	52.25	10.12	5.91	5.91	—	60.58	35.41
Total	927.25	329.42	35.52	58.68	6.33	68.80	7.41	470.35	50.72
P.S. Salanpur									
Mahishmura	143.53	8.09	5.63	93.07	64.84	1.62	1.12	40.75	28.39
Malikola	125.33	—	—	115.34	92.02	6.07	4.84	3.92	3.12
Dharmma	108.35	—	—	58.68	54.15	23.47	21.66	26.02	0.24
Talbiriya	143.96	—	—	135.17	93.81	8.79	6.01	—	—
Ethora	515.14	—	—	242.81	47.07	10.12	1.96	262.61	50.93
Total	1026.71	8.09	0.78	645.07	62.83	50.04	4.88	333.48	32.48
P.S. Neturia (Purulia)									
Mahisnadi	296.22	—	—	87.01	29.37	85.79	28.96	123.42	41.66
Kelyasota	299.22	—	—	87.41	29.05	85.79	28.96	126.03	42.54
Total	595.44	—	—	174.42	29.29	171.58	28.81	249.45	41.89
Total study area	8048.07	786.34	9.75	3414.25	42.41	865.28	10.75	2982.08	37.05

north-western corner of this area, particularly in the Kulti police station, the slope is steep to moderate, it ranges between 2°-2.5° and above. Basement complex rocks such as granite and gneissic outcrops are found here. Because of this, slope is higher than the other areas. In the south-western fringe of this area, where Damodar river is flowing, moderate to high sloping land is found. In Purulia district, near Paraballiya there is high sloping land with spurs and valleys forming a highly rugged terrain. Slope of this area is near 2°. In Dishergarh, at valley side of the Damodar, highly sloping land with undulating rocky outcrop has developed mainly on Gondwana formation.

Rest of the areas have 1.5°-2° slope with moderate, gentle gradient terrain. Overall, the region is lowlying, undulating erosional surface.

Cultivated Land

The discussion of slope and cultivated landuse pattern is based mainly on the field observations and block-wise statistics. In the Damodar-Dhajmma interfluve mainly five types of landuse categories are prominent. They are: forest land, irrigated land, unirrigated land, cultivable waste land and area not available for cultivation. Forest lands are out of question for cultivation. Irrigated and unirrigated lands are cultivated lands, cultivable waste land can be converted into cultivated land if land potentialities prevail there. The land which is not available for cultivation is used mainly for settlement or road construction.

(i) Block-wise land utilisation

It is evident from the Table 1 that in 1991, there was no trace of natural forest. During this period, 51.17% of the study area was covered by cultivated land. Of this cultivated land, 9.76% was irrigated and 42.4% was unirrigated. Cultivable waste land was 10.75%, and 37.05% area was covered by uncultivable land (area not available for cultivation). In the Kulti block, percentage of irrigated area was very low (0.33%). Most of the cultivated land was unirrigated (55.7%) within this study area. Asansol police station covers the highest percentage of irrigated land (Figure 3) but in the areas of Salanpur police station, percentage of irrigated land is very low (0.78).There is no change from 1981 to 1991.

(ii) Field observations on landuse

Kulutara village under the study area was visited. The major part of the village settlement is situated in the northern side of GT Road and southern part of Kulutara village is mainly used for cultivation. Total cultivated area of Kulutara village (Table 1) is 178.91 hectare (78.91 hectare is unirrigated). In between cultivated lands, there are some barren lands but this area is not well suited for agriculture because of topographic variation. This area is a part of Chotanagpur Plateau. So, rolling and undulating topography is predominant. Steep and abrupt slope is conspicuous in this village. Irrigational facility is nil. Due to this topographic variation, terracing agriculture is practised here. Some villagers said that the agriculture of this area mainly depended upon rain water.

Another village was also visited which is an industrial area in Kulti. Urban agglomeration is conspicuous than the other areas within this study. The principal economic activity of this region is industrial work but some agricultural fields within this industrial belt were observed. Though the people are engaged in industrial work, they practise agriculture as extra activity. We had a misconception that in industrial areas, agriculture practices are not being carried out. But in Kulti area, this conception has proved to be wrong because we saw an extensive agricultural field. It is important to note that census reports do not reveal the agricultural area but cultivated lands are there and persons are interested in cultivation also.

The Dishergarh area which is mainly coal mining area was also visited. The main economic activity of this region is mining and associated local industries. Besides these, agriculture is practised here. Many people are engaged in subsistence agriculture. Mainly *aman* paddy and wheat are cultivated here.

Figure 3

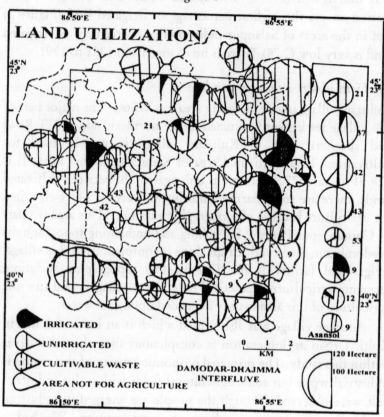

Correlation between Slope and Agricultural Land

Slope is the dominant factor for the development of agriculture. Therefore, the authors tried to correlate agricultural land with slope categories. With the help of bivariate analysis the relationship between slope category and agricultural land (irrigated and non-irrigated) has been shown.

It becomes evident that as the degree of slope increases (Figure 4, Table 2), the agricultural land decreases. The relation is negative. Here co-efficient of correlation is --0.14. The following observations have been made:

Figure 4
Bi-variate Analysis between Slpe and Agricultural Land

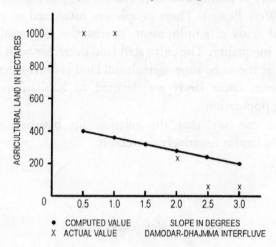

COMPUTED VALUE   SLOPE IN DEGREES
X ACTUAL VALUE   DAMODAR-DHAJMMA INTERFLUVE

Table 2
Linear Regression Equation Between Slope and Agricultural Land

Slope (X)	Agricultural land (Y) in hectare	X^2	XY	$YC=a+bX$
0.5°	988.18	0.25	494.09	401.14
1.0°	1006.32	1.00	1006.32	354.98
1.5°	362.82	2.25	544.23	308.82
2.0°	202.15	4.00	404.30	262.66
2 5°	35.00	6.25	87.50	216.51
3.0°	20.00	9.00	60.00	170.35
$\Sigma x = 10.5$	$\Sigma y = 1714.47$	$\Sigma x^2 = 22.75$	$\Sigma xy = 2596.44$	

The major cultivated area lies mainly on lower gradient (between 0.5°-1.5°). The areas where slope exceeds 1.5°-3°, the agricultural land under cultivation is very low.

Agricultural activities though poor in status, some observations and statistical data give a different picture. The land under agriculture is nearly 1000 hectares, where slope ranges from

0.5° to 1.5°, but the cultivated area of this region is 2500 hectares. A large number of immigrants have come from other states and other parts of West Bengal. These people are interested in agriculture because of crisis of employment, overburden of population and economic inequality. The cultivated land decreases with increasing slope but at the same time agricultural land is decreasing at a very rapid rate as these lands are devoted to habitations to settle increasing population.

So we can say that the relationship between slope and agricultural land is negatively correlated.

References

Basu, M. (1998): Terrains and the Development of Agriculture in the Madhupur Erosion Surface in Bihar, unpublished Ph.D. Thesis, Burdwan University.

Bose, A. (1980): *Statistics.*, Calcutta Book House, pp. 35-75.

Prasad, N. and Gupta, S.N.P. (1981): Streamline Surfaces of the Barakar Basin, *Perspective in Geomorphology*, Vol. 4, Edited by H.S. Sharma, New Delhi, pp. 109-130.

Prasad, N. and Khadse, N.H. (1990): Correlation between Slope and General and Agriculture Landuse in the Pus Dam Catchment of Akola and Yavatmal Districts, Maharashtra, *Geographical Perspective*, Vol. 4, No. 1,2.

14

Soil in the Hazaribagh Plateau: A Case Study in Land Degradation and Conservational Planning

Onkar Prasad

Next to air and water, soil is of vital significance for the existence of all types of life on earth. "It is regarded by many as the major element of land. Soil characteristics to a large extent determine some basic land qualities such as the ability to supply water to plants, streams and wells, etc.; nourishment for plants and animals; materials for construction purposes; and mechanical support for plants, animals, human beings and the structures they put up" (Joseph Uyanga, 1991). It is now realized that in any comprehensive regional planning programme, there is need for a regional soil suitability study which maps the distribution of the various types of soil, analyses their properties and interprets these for landuse.

The Study Area

The Hazaribagh Plateau (between latitudes 23°36' to 24°48' N and longitudes 84°27' to 86°35' E), with an area of 15,129.5 km², forms the north-eastern part of Chhotanagpur Highlands of the state of Bihar (Prasad, 1983).

Physiography

The landscape of the study area expresses itself in different surface forms which range from granite-gneiss domes, projecting over peneplained surface, to peaks, hills, spurs and scarps highly dissected and diversified by streams. The general elevation of the area varies between 250 m and 600 m above mean sea level. In fact, the broad and undulating plateau consist of two distinct erosional surfaces -- the Upper Hazaribagh Plateau surface and the Lower Hazaribagh Plateau surface, having an average elevation of 600 m and 400 m respectively. The former has length (east-west) of 60 km and width (north-south) of 25 km, the area of 956.94 km² being 6.37% of the total study area. On the other hand, the latter covers 93.63% of the total study area, has been divided into four sub-regions: the Kodarma Plateau, the Chatra Plateau, the Bagodar Upland and the Northern Dissected Fringe (Figure 1B). The descent from upper surface to the lower surface is not gradual but rather abrupt in the form of narrow and steep slopes or scarps, locally known as 'ghats'. Further, the relative relief (100 to 400 m) of the upper surface closely resembles that of lower surface but this is not so in the case of their escarpments. The latter differences arise most probably out of the differences in rock-resistance to erosion in the two cases. In fact, the southern escarpment of the upper surface is steeper (above 6°), than the northern escarpment of the lower surface (below 6°), the latter also being more dissected. The slope of the upper surface (below 1°) is also gentler than the lower surface (1°-2°).

Structurally, the study area is an ancient landmass; it has perhaps never gone under the sea and has been, therefore, subjected to long-continued soil erosion. As regards the surface drainage, the Hazaribagh plateau is a part of Ganga basin and finds its outlet through the Ganga system into the Bay of Bengal. According to the drainage characteristics, the drainage of the area can be grouped into six main systems: the Damodar, the Sakari, the Mohane, the Amount, the Lilajan, and the Morhar. Although only a part of the Damodar system lies in the area yet it constitutes the most important drainage system with a drainage area of about 10,580.75

...of 60 23' of the south... area. The Damodar valley has been carved out in rocks of varying resistances, ranges from highly resistant crystalline rocks of the Archean to the less resistant sedimentary rocks of the Gondwana period.

Rainfall

In conformity with the characteristic nature of the monsoon rainfall, the rainfall distribution in the area establishes marked variations with respect to other factors. The humidity decreases with a decrease in rainfall amount. More than 90% of the total annual rainfall occurs during the rainy season. The increase in a deficit of precipitation with water being the excess in the drier areas. Higher humidity... ...decreases gradually from east to west... ...part of the territory mainly...

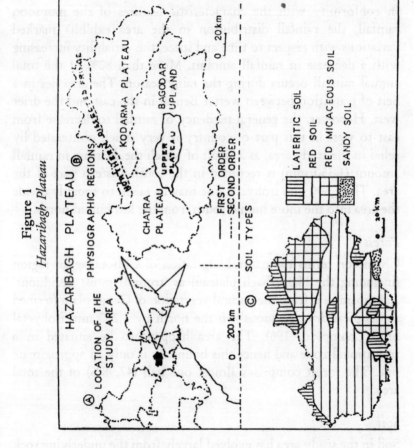

Figure 1
Hazaribagh Plateau

Ⓐ LOCATION OF THE STUDY AREA

Ⓑ PHYSIOGRAPHIC REGIONS

HAZARIBAGH PLATEAU

KODARMA PLATEAU
BAGODAR UPLAND
UPPER PLATEAU
CHATRA PLATEAU
NORTHERN DISSECTED FRINGE

—— FIRST ORDER
- - - SECOND ORDER

0   20 km

Ⓒ SOIL TYPES

LATERITIC SOIL
RED SOIL
RED MICACEOUS SOIL
SANDY SOIL

0   20 km

Soil

Soil in the study area has evolved largely from the underlying rock. In... through geological time. The stream valley bottoms are the only sites covered by transported alluvial soils. At several places, excessive water erosion has produced deep gashes in the soil cover. The parent rocks thus exposed have often subsequently been covered afresh by rock materials washed down from higher levels.

km² or 60.82% of the study area. The Damodar valley has been carved out in rocks of varying resistances, ranging from highly resistant crystalline rocks of the Archean to the less resistant sedimentary rocks of the Gondwana period.

Rainfall

In conformity with the characteristic features of the monsoon rainfall, the rainfall distribution in the area exhibits marked variations with respect to time and space, the variability increasing with a decrease in rainfall amount. More than 80% of the total annual rainfall occurs during the rainy season. The area lies in a belt of transition between wetter Bengal in the east and the drier west. However, the general tendency of rainfall to decrease from east to west in this part of country is very much obliterated by relief in the study area, as a result of which the decrease in rainfall amount (1,350 mm) is recorded in the south-western part of the area. The 1,300 mm isohyetal line may be taken to roughly divide the area into the more humid south from the less humid north.

Forest

Champion terms the natural vegetation of Chhotanagpur region (including the Hazaribagh plateau) as 'tropical moist deciduous' and separates it from the natural vegetation of Ganga plain (termed as 'tropical dry deciduous') in the north by 1,200 mm isohyetal line (Champion, 196). The area happens to be situated in a transitional zone and hence this boundary is only an approximate one. The forest comprises almost one half (47.98%) of the total area.

Soil

Soil in the study area has evolved largely from the underlying rock materials through geological time. The stream valley bottoms are the only sites covered by transported alluvial soils. At several places, excessive water erosion has produced deep gashes in the soil cover. The parent rocks thus exposed have often subsequently been covered afresh by rock materials washed down from higher levels, for instance, in the valleys of Barsoti and Jamunia rivers. In the greater part of the study area, however, the soil is composed of infertile debris, much of which is still in the process of disintegration. In general, the lowland areas (locally known as 'dons') of the undulating surfaces of the plateau are comparatively more fertile. The uplands (locally known as 'Tanrs') are well drained, gently to moderately sloping, and moderately to roughly eroded. The lowlands, on the other hand, are moderately to roughly drained, gently sloping and generally bunded. Soil erosion is thus largely confined to the uplands.

Soil Types

The spatial variations in the soil characteristics in the study area are largely influenced by the parent rock and by the extent and nature of weathering. Moreover, local changes in land elevation and water availability bring about consequential change in soil catena. The soil of the study area, primarily residual, has been divided into four major soil groups (Figure 1C):

1. Lateritic soil
2. Red soil
3. Red micaceous soil
4. Sandy soil

Lateritic Soil

The term 'laterite' generally denotes a soil formed *in situ* by the leaching out of the mineral bases and much of the silica from the original rock, leaving a residue rich in alumina. Normally, these soils are derived from argillaceous material impregnated with iron-peroxide and mottled with various hints of brown, red and yellow, while a considerable proportion sometimes consists simply of white clay. The soil usually hardens on drying up. Soils derived from the laterites, rich in hydrates of alumina and formed under high seasonal rainfall conditions, get their silica removed in solution through the process of leaching, leaving behind hydroxides of alumina. It is often difficult to fix common chemical characteristics of laterite soils.

In the study area, the lateritic soil is observed on the higher plateau surface or summits of hills like Parasnath, Chenhri, Lugu Pahar, Nawadih, and in the south-central part of the Mohane-Damodar divide. Further, some parts of Siwani-Barsoti divide, Kewata-Noni divide and hills of the northern dissected fringe are capped by these soils. At present the soils are extensively found between elevation of 400 m to over 1,200 m. However, they originally occurred at elevations of 700-1,300 m and 400-600 m, representing two prominent erosion surfaces in the area, progressively breached by water dissection. The eroded laterite has been deposited at lower elevations, mainly between 350 and 500 m, as secondary lateritic sheet wash material. It overlies the disintegrated laterite gravels and varies in thickness from 1 to 10 m. The laterite gravels, boulders and cobbles cap the hill tops and the uplands, for example, the Parasnath hill. The hill, formed by the process of circumdenudation, has a much extensive copping of laterite in the recent geological past.

The lateritic soil is generally poor in lime and magnesia and its phosphorous, nitrogen and potash contents too are quite low. Having developed under monsoonal climatic regime, it has undergone thorough leaching so that its exchangeable bases have been removed. Thus, the soil is of low fertility and can support crops like 'kodo' (an inferior variety of rice), 'surguja' (oilseed), 'kulthi' and 'bodi' (both pulses).

Red Soil

It is an extensively distributed soil covering 62.34% of the study area. It is derived from the ancient crystalline and metamorphic rocks -- granites, gneisses and schists, with subordinate rocks rich in ferromagnesium minerals. Its texture and colour vary from the eastern to western part of the study area. The soils of the eastern part are light to medium in texture and red to yellow and light gray in colour. The soil is locally known as 'Lalki matti,' but at a few places some dark gray varieties also occur, known as 'Kewal'. The latter is the most fertile and is well suited for paddy cultivation.

The soil covers most of the intermediate slopes of river valleys and of uplands. The upper valley of Damodar and the Chatra plateau is also capped with this soil (Figure 1C). At some places, the soil merges into the lateritic soils. The commonest form of the red soil is a sandy clay of the uplands to the rich, deep, fertile loams at lower levels. Thus, crops requiring less water (millet or bajra, arhar and surguja) grow on the former and rice, maize, wheat and gram on the latter.

Red Micaceous Soil

The mica belt of the study area, mainly the Kodarma plateau and the areas around Mandu and Churchu of Bagodar upland are covered by this type of soil. It is a light pinkish soil on the uplands and a yellow to light yellow on the lowlands. Soils are loamy sand in texture, with high proportion of mica particles. Although the soils are fertile, yet in the absence of sufficient water, they are not able to produce a good harvest. The cropping pattern in the soil largely resembles that of the previous soil.

Sandy Soil

It is the least productive soil of the study area, and is locally known as 'negra' and 'rehra'. It is poor in nitrogen, phosphorous and humus but has moderate amounts of potassium and lime. It is moderately deep and the colour varies from reddish-yellow to greyish-yellow. The soil is mostly found in the south-eastern part of the study area, mainly in the stream flood-plains.

Fertility and Productivity of Soil

The fertility of soil depends upon the biological, chemical and physical properties of the soil (Gruickshank, 1972). The soils of the uplands are coarse textured (loamy sand or sandy loam) and are generally shallow to moderately deep, slightly acidic in reaction and relatively poor in fertility. They are mainly cultivated for dry crops. On the other hand, the soils of the lowlands are moderate to fine textured, with low permeability but relatively high water holding capacity, neutral to slightly alkaline in reaction and possessing a higher fertility. However, Figure 2 (based on the surveyed data collected from D.V.C.) reveals some fertility characteristics.

Nitrogen

Almost all the agriculturally significant soils in the study area are deficient in nitrogen -- the most important nutrient (Figure 2A). This deficiency could be made up by utilising cow-dung as compost instead of burning it as fuel. The nitrogen deficiency has been aggravated from soil-erosion in the uplands and from water-logging in the lowlands.

Phosphorous

Except Pratappur, Nawadih and Gumia blocks, almost all areas have low to moderate phosphorous content in the soil (Figure 2B). The deficiency is aggravated due to causes responsible for nitrogen deficiency.

Potassium

The potassium content in soil is relatively higher in the study area in comparison to nitrogen and phosphorous contents. Mostly the outer fringes of the study area like Barkagaon, Pratappur, Deori, Nawadih and Gumia are sufficiently rich in potassium (Figure 2C).

Acidity and Alkalinity

Most of the upland soils are acidic in nature, for example, in Ichak, Dumari, Hazaribagh, Mandu, Giridih, Jainagar and Barhi, whereas the lowland soils in these blocks are slightly acidic to neutral in reaction. In the rest of the areas, the soils are from slightly alkaline to neutral (Figure 2D).

Soil Erosion

The highly uneven topography, alongwith a high incidence of intensive rain-storms, accelerate the rate of run-off from rainfall, resulting in a high rate of soil-erosion. Low soil permeability, deforestation, overgrazing and upland cultivation act as

Figure 2

Hazaribagh Plateau: Soil Characteristics

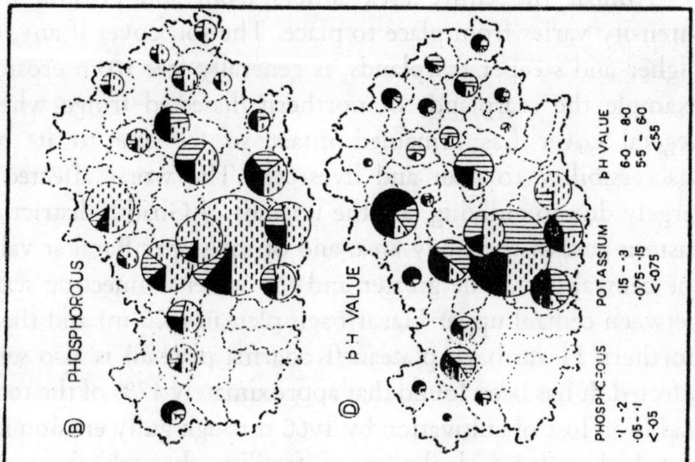

contributory factors. Sheet erosion is a common feature, causing heavy annual loss of valuable top soil. Gully erosion too is quite prominent, resulting in the conversion of agricultural land into ravine land in the land degradational processes.

Almost the entire area suffers from soil erosion but the intensity varies from place to place. The soil cover if any, on the higher and steeper scarplands, is generally free from erosion, for example the scarplands of northern dissected fringe where the vegetal cover has remained intact so far due to its general inaccessibility to man and livestock. The worst affected areas, largely due to gullying, are the uplands of Giridih district, in the eastern part of the study area, and of the upper Barakar valley, in the central part. The greater and the wider connecting scarpland between central upper Hazaribagh plateau (600 m) and the lower northern Hazaribagh plateau (Kodarma plateau) is also seriously affected. It has been found that approximately 17% of the total area had been lost of cultivation by 1960 through gully erosion and the rest had suffered depletion of fertility through sheet erosion (Bhattacharya, 1964). This has also resulted in siltation in reservoirs, and riverbeds, thereby reducing water supply and increasing flood frequency. A fall in the storage capacity of reservoirs has also resulted in a fall in the water-table of the surrounding areas and drying up of wells in extreme cases. The commonly utilised soil conservation practices are contour-ploughing and terrace cultivation on slopes and construction of check-bunds or 'bandhis'.

Soil Conservation and Planning Proposals

The 'Soil Conservation Development' of the Damodar Valley Corporation (DVC) has adopted soil conservation measures with a views to mitigate the problem of sedimentation in the reservoirs and the Damodar river. Similar measures have also been adopted by the state government, for the reservoirs managed by them. An active participation of the population is a pre-requisite in the successful implementation of these measures, initiated on a large scale from 1955. In fact, an integrated approach is required for the utilisation, maintenance and conservation of soil resources through proper coordination between the soil conservation department on the one hand and others like the irrigation, forest and animal husbandry departments on the other.

Soil erosion and low fertility are the main problems related to use of soil resources. As regards checking soil erosion, special attention needs to be given to the catchments of upper Barakar and Jamunia river which are severely affected. The state government forest department has been undertaking reforestation programmes and checking overgrazing and other destructive landuse practices in time with the objectives of the National Forest Policy adopted for the country. Proper forest management practices need to be urgently applied in the severely affected areas of soil erosion in the Hazaribagh plateau. The government agencies should also involve tribals and other unemployed people in the soil conservation and reforestation programmes. This will not only lessen the local unemployment problem but also develop conservation consciousness amongst the inhabitants. There is need to move from general exploratory approach to a more objective and systematic method of soil study that places emphasis on direct observation, measurement and analysis of soil properties to reduce the land degradation.

References

Bhattacharya, J.P. 1964. *Study of Soil Conservation Programme for Agricultural Land*, Bombay, p. 208.

Gruickshank, J.C. 1972. *Soil Geography* David C. Charles, Newton Abbot, p. 202.

Joseph, U. 1991. *Soil Survey and Mapping: A Development Planning Perspective in Nigeria*, East West Geographers, I.G.S., Munger, Vol. 2, No. 2, pp. 8-17.

Prasad, O. 1983. *An Appraisal of Resources for Economic Development in the Hazaribagh Plateau*, Ph.D. thesis, B.H.U. Varanasi, India.

15

Degraded Land: Management and Traditional Wisdom (A Case Study of North Sikkim Himalaya)

S. Gangopadhyay

The North Sikkim Himalaya is conspicuous by its lofty height, steep slope, splendid natural beauty along with blissful tranquillity, sharp contrast in environmental situations within very short distance, and extremity in environmental set-up. It also gains prominence in low density of population, and dependence on degraded land by non-agricultural uses. The location of third highest peak Kanchenjengha, I-shaped gorge of Tista valley due to structural weakness, change in production system from terrace and slope cultivation to pastoralism with gaining altitude as a way of management in response to quality of land and nature, has brought about the varieties in human adaptation. Land becomes degraded or poor due to steep slope and height, severity of winter, and poor rainfall and high altitude in the North Sikkim Himalaya. It is the ideal situation for studying poor quality of land, i.e., degraded land. Majority of its land (greater than 50% of total land) lies at an elevation of about 3000 m, which is practically not fit for cultivation. The area of Dombang and north of it in Lachung valley; Muguthang, north of confluence of Lohonok and Pokechu or Latong-Denga; sub- area of southern part of Lachen valley imposes a great limitation on continuation of agricultural functions due to severity of winter marked by nearness of snowline, rocky and rugged terrain. Therefore, pastoralism marked with seasonal migration to utilise the forage and land available conveniently for livestock in the respective seasons has been selected as the best possible way of management inspite of it having all kinds of limitations. Similarly, the upper Zongu lying south of Chungthang has restricted cultivation to normal extent mainly due to inaccessibility and steep slope. Shipgyer is one such place where slash and burn method of cultivation replaces normal cultivation as the best possible way to combat with negative effects of steep slope.

Thus, the paper tries to critically analyse the management procedure resorted to by the local people of the northernmost part of North Sikkim district, i.e., uppermost part of Tista valley. It is also shown how the way of management combats with extremity of environmental situation for the purpose of turning the degraded land into a resource-generating instrument to the best possible extent.

Degraded Land

Land becomes degraded due to cultural factors such as over-use of land through high intensity of agricultural activity, much use of water from the ground or surface water, use of high-yielding variety seeds, chemical fertilizers, pesticides, etc., addition of industrial wastes in the land, mining and quarrying for the use as energy or for development purpose, deforestation due to development, overgrazing, etc. But, the physical factors such as unfavourable position of physiography marked by rugged terrain, steep slope, lofty height with ramifying outer-spur, sharp contrast in climatic conditions within very short distance and prolonged winter, snow cover due to nearness of snowline, landslides, soil erosion, water-logged conditions, sand deposit, salinity and alkanity make the land poor and unfit for cultivation. Thus, the degradation is not the result of a single factor, but is dependent on multiple factors. Inspite of having synonymous character with wasteland in terms of its unsuitability for cultivation or little potentiality in productivity, controversy arises when these two terms 'waste land' and 'degraded land' are said as having same meaning (Sagwal, 1991). However, in respect of identifying the factors responsible for degradation, concerned departments/ experts put them in physical and cultural lines with a few differences in marking the level of factors. The concerned departments/experts are Society of Wasteland Development, National Remote Sensing Agency, National Wasteland Development Board, and Shegal Jawahar, L. (1986) who deserve special mention in this regard.

Considering all the points made by different departments/ boards/experts, it has been understood that degradation means that land which is unfit or becomes unfit in giving optimum returns due to natural limitations, developmental or cultural factors, is either degraded or undergoing degradation irrespective of its status of not taken up for cultivation at all or taken up for cultivation once or left out for the current year/following years, stony waste, forest/pasture lands, barren rocky areas, snow/glacier covered area, shifting cultivation areas, etc. Coming to the situation in the area of North Sikkim Himalaya, it may be said that height and slope along with severity of winter greatly contributed to land degradation, either in the village or outside the village. Now, for the purpose of elaboration of subject matter, the way of management and application of traditional wisdom in using degraded land, culturable wasteland and land not fit for cultivation has been taken into consideration. The area lying above 5000 m contour line is not taken up for discussion, though it is degraded technically, because its use has not become possible by the local people due to its inaccessibility.

Extent and Limit of Degraded Land

Before going into detail over the management procedure of degraded land through traditional wisdom, it is better to have an idea on extent of degraded land and its limit at macro and micro level of North Sikkim Himalaya. Normally, degraded lands are degraded forest lands, shifting cultivated land, degraded pasture/grazing land, gentle slope with undulating flat terrain having meadows and grasses, etc. In North Sikkim Himalaya, there are some areas lying at a height of 3000-5000 m in the northern most part of the state where grasses come out at the advent of summer. The area is used by the local people through livestock grazing in the form of transhumance system. These are far away from natural habitation area. But, above 5000 m, the area is extremely high mountainous zone and inaccessible. So, it has turned into permanent degraded land in non-usable form. Similarly, this type of degraded land is found within the limit of habitation zone mainly because of peculiar physiographic position. The present discussion centres around only degraded lands which are in the form of its use, irrespective of its position in or outside habitation zone.

It is found from the range of height of North Sikkim Himalaya that more than 50% of land lies above 4200 m contour line, very near to permanent snowline lying with 4600 m contour line. Above it, the zone is very difficult for human activity due to severe winter and rugged terrain. The altitude zone of North Sikkim is as follows:

Altitudinal Zone (North Sikkim)

Altitudinal zone (in m)	Percentage of area of total land of North Sikkim district
> 1800	5%
3000 - 1800	13%
4200 - 3000	24%
Above 4200	58%

Note: Total area of North Sikkim district is 4226 km^2.

At micro level, it is seen that Lachen and Lachung have two types of settlements -- permanent and temporary, which correspond roughly to two altitudinal zones -- upper and lower part. But, the limit of having settlement in the upper and lower part of the two villages differs and is guided by the seasonal suitability in two villages. In case of Lachen, the upper part varies between 3000-4000m and is known as summer settlement, though livestock moves little higher of it, while the lower part mainly known as permanent settlement is confined to 2000-3000 m. It is also known as winter settlement. In case of Lachung, notion of permanence or temporary status is guided mainly by agricultural activity as against livestock movement at Lachen (because pastoralism is main livelihood). However, at Lachung, permanent settlement (all weather settlement), i.e., Bicchu, Singring, Thomchi, Sorchuk lies between 2500-2700m, while temporary settlement (Leema, Khedum, Maltim) lies between 1400-2200m, only during agricultural operations March to June. In respect of Shipgyer village, the height varies mainly from 1350-2000m though a few portions touch even 3000 m.

But, this village is noted for its ruggedness. The ratio between horizontal and vertical fall is 100:4. The same ratio and cross-profile was taken above Lachen settlement (3000 m). But, it is reverse in case of Lachung valley. The upper portion is relatively more rugged in comparison to lower counterpart in Lachung valley.

Causes of Being Degraded Land

Apart from slope and increasing height within short distance, severity of winter marked by lowermost temperature (even below freezing point) due to nearness of permanent snowline at 4600 m height, decreasing rainfall with gaining altitude, and higher acidity make the land unfit for cultivation and thus degraded land. The table below reflects the decreasing rainfall with gaining altitude.

Rainfall (in mm)

Place	Chungthang (1500 m)	Lachung (2624 m)	Lachen (3000 m)	Thangu (3658 m)
Average annual rainfall	2775	1898	1485	531

Note: Soil is mostly acidic in nature.

Thus, the North-Sikkim Himalaya has more than 50% land lying above 4200 m which is almost degraded land. But, in such areas, viz., Lhonok, Dongkong, Muguthang, Yumthang, extensive grassy meadows are found with the advent of summer. During this time, the land becomes unfit for grazing in summers. Similarly, in Shipgyer village, the area is mostly degraded due to ruggedness in terrain except some narrow flat land available between slopes, which can be used for cultivation to a little extent.

Way of Management and Traditional Wisdom

The meagre agricultural land in the higher part of North Sikkim Himalaya has led to livestock grazing through transhumance systems. It is found from the data that Lachen has about one sq km of cultivated land while Lachung has about 5 sq km of cultivated land. Naturally, Lachen is bound to take resort to livestock grazing even in the degraded land at high altitude, outside or inside the village. The use of forage is regulated by the change of season. Thus, livestock moves upward/downward and use the grasses grown in the meadows. In addition to livestock grazing, they collect fuelwood/medicine from forest which grows in the higher altitude. Again, shifting cultivation in 'Sosyum' (temporary field) and social forestry have been taken as a way of management in the Shipgyer village.

Pastoralism through transhumance is more prominent at Lachen due to available grassy meadows even above 3000 m. There are about five sub-areas, namely, Dongkong, Muguthang, Thangu-Taling, Samdong-Yathang-Kalep and Raufu-Yunga-Sacham, lying at different altitudinal tiers varying from 3500-5000m. Similarly, livestock moves about in the Chaten-Lachen-Zemateng and Laten-Rabong sub-areas lying between 2100-2600m. Again, in Lachung, high altitude grazing is not prevalent due to scarcity of grazing land at higher elevation and cessation of animal movement suited to high altitude after the close of border with Tibet in 1962. However, meagre grassy land is found at Yumthang-Mimisamdong-Jadung-Dongkhela and Sebo-Phekoching-Chumbakha lying above 3000 m at Lachung. The dearth of high altitude suited animals at Lachung (due to closing of border with Tibet), restricts high altitude grazing in spite of having few grassy lands in and around Yumthang. But, it is to be noted that grazing at pasture land and cultivated land is quite common at Maltim-Yaksha sub-area (2000-3500m). During heavy winter, livestock moves further down to Toong-Naga sub-area (1200-1800 m).

Ponies, even yak, were used for transportation in trade activities in Tibet before the close of border. However, cattle is reared in the hope of getting milk, and woolen products. Meat is also consumed. Thus, it is seen that forage grown in different ecological sub-areas are used inspite of having shortage of grazing land and change in type of animals after border close. The following tables are presented here in order to explain the level of livestock grazing.

Livestock Population at Lachen and Lachung (No. of cattle per household)

	Yak	Sheep	Goat	Cattle	Horse and Ponies
Lachen	17	29	5	4	3
Lachung	5	1	-	7	3

High-altitude Livestock at Lachen (No. of cattle per household)

Tibetan Yak	Tibetan Sheep	Tibetan Goat	Total
74	158	32	264

Note: Based on 15% sampling of total households whose livestock graze exclusively above Thangu.

In this connection it is also to be mentioned that the Lachenpas know the use of high altitudinal area (3500-4200 m) lying in and around Thangu and Kalep. They utilise it by selecting potato cultivation in summer. During this time, land becomes ice-free and enriched with cattle-dung left out during livestock movement. It is also a fact that 78% of total cultivated land is under potato cultivation of high altitude area imposed by limitations of freezing temperature for major time of the year, and slope. However, this kind of cultivation is very little in relation to total geographical land.

The general landuse at Shipgyer village clearly indicates that 69% of total land, i.e., 16 sq km is not available for cultivation due to high altitude and steep slope. It is degraded land in non-usable form. Again, 15% of total land except a small portion (38 hectares) is under degraded due to steep, narrow, stony land and poor soil development. This type of land is called 'Sosyum' in Lepcha language. Narrow, flat land present between the hills is locally known as 'Sing' (permanent land) which allows cultivation of maize, wheat, soyabean. In order to restore the fertility, the land is kept fallow for four years on an average and cultivation is allowed only for a year or two. About 72% of the total farm land is left fallow every year due to poor fertility. In such kind of degraded land, crops grown are maize (21% of total cultivated land), followed by *kodo* and *fapar* (millet variety crops). Again in order to combat with resistance offered by such degraded land, maize is sowed in the holes at an interval of half metres, while *kodo* and *fapar* are sown broadcast. But, in contrast, maize is broadcast in permanent field and kodo is transplanted. It is striking to note that *tokmorjo* and *chobyojo* are the dry paddy varieties grown in the rugged terrain.

Another striking feature is that large trees are cut from a certain height, while the smaller ones are cut from the ground. Next, the trunks of the trees are laid along the contour lines. The varieties cut for fuelwood are *Kartoosh, Mawa, Flant, Urtish, Chillawani*.

The trees and bushes get dried for about one month and burning takes places by 15th March. The spades are used for digging the soil. Sticks and *kantas* are used to dig soil, particularly in stony area and in slope. Sickles are used in harvesting. Ashes of burnt trees and bushes are used as manure in temporary fields. Therefore, it is seen that Lepchas utilise such degraded land through selection of crops and crop variety, clothing appropriate method in tune with suitability of land. They are even aware of soil erosion, and it is indicated by cutting trees at certain height.

Conclusion

In view of the discussion, it is observed that environmental limitations and constraints have a great bearing on making land degraded and poor for cultivation. Accordingly, mode of production system, i.e., livestock grazing through transhumance system, high altitude adapted livestock grazing, utilisation of forest resources, slope cultivation allowing restoration of fertility of poor land during regeneration period of four years, potato cultivation in high altitude in Lachen during summer, are in tune with the condition of degraded land. Understanding terrain and soil condition, seasonal change, people regulated movement of livestock at different tiers of height, and selecting high altitude livestock, is the only way to utilise forages available at a height of 3000-5000 m. Even agricultural land of the two valleys of the higher reaches is enriched with cattle-dung, when livestock moves downward or upward during seasonal change. Apart from this, crop and varieties, technique of sowing the crop seed in the degraded land in the form of putting the seed into hole or broadcasting, use of digging stick and *kanta* (used for removing stones and pebbles), fixation of regeneration period, use of forest species as fuelwood grown during regeneration period, cutting the trees at certain height to prevent soil erosion before the commencement of cultivation, laying of fallen trees along the contour line to regulate water and to prevent soil erosion, are the striking features which adapt with the condition of degraded land at grassroot level for the purpose of making the land productive as far as possible. The use of forest species as fuelwood or domestic requirement or medicinal plant is also an indication of dependence on forest resources for sustainable purpose. Thus, it may be said that the use of some portion of degraded land by using traditional ways has mitigated the problem arising out of paucity of land for agriculture, horticulture, and plantation.

References

Banerjee, Bireswar. 1992. In *Development and Ecology*, Mehdi Raja (ed.), Rawat Publications, Jaipur/New Delhi.

Bhasin, M.K. and Veena Bhasin. 1995. Sikkim Himalayas: Ecology and Resource Development, Kamla Raj Enterprises, New Delhi.

Bose, Saradindu. 1966. The Bhot of Northern Sikkim, *Man in India*, Vol. 46, No. 2.

Gangopadhyay, S. 1983. Human Adaptation in the Sikkim Himalaya: Some Observations, paper presented at the workshop 'Human Adaptation in the Extreme Climate' on the 27th June, 1983, at Anthropological Survey of India, Calcutta.

Gangopadhyay, S. 1985. Transhumance in the North Sikkim Himalaya: A Way of Environment, in 'Impact of Development on Environment', (ed.), Vol. 1, Bireswar Banerjee, Geographical Society of India, Calcutta.

Gangopadhyay, S. 1988. Nature of Human Adaptation in the Tista Valley of Sikkim: A Geographical Analysis, Ph.D thesis, Calcutta University.

Gangopadhyay, S. 1989. Human Adaptation in a Changing Situation: A Case Study of Lachung, a Village in the Sikkim Himalayas, in 'Acculturation and Social Change in India' (ed), S.C. Panchbhai, M.K.A. Siddiqui, Anthropological Survey of India, Calcutta.

Sagwal, S.S. 1991. 'Wasteland in Kashmir: Some Remedies', in 'Kashmir: Ecology and Environment' (ed.) Chadha, S.K., Mittal Publications, New Delhi.

Sehgal, Jawahar, L. 1986. Degraded Lands of Arid and Semi-Arid Regions of India and their Land-use Planning, paper presented at Summer Institute on 'Waste lands in arid and semi-arid zones and technologies for their improved utilization', 10-13th June, 1986, CAZRI, Jodhpur, Rajasthan.

16

Nature and Problems of Premature Reclamation in Sundarban Delta

S.N. Chatterjee

Sundarban which is located in the district of North and South 24 Parganas of West Bengal, falls a little south of Tropic of Cancer, between the latitudes 21°32' and 22°40' N and longitude 88°05' and 89°10' E longitudes. The total area of Sundarban region in West Bengal is 8,373 km², bounded by Dampier-Hodges Line as per survey and demarcation done during 19th century. The greater part of Sundarban falls in Bangladesh and only one-third lies in West Bengal. In West Bengal, it is bounded in the west by the river Hooghly, in the east by the Raimangal river and in the south by the Bay of Bengal. The entire area belongs to active delta zone of the river Bhagirathi (Ganges) and is crisscrossed by innumerable rivers and creeks, viz., Ichamati, Kalindi, Saptamukhi, Thakuran, Matla, Bidyadhari and Raimangal, etc. The area is constantly growing with the deposition of river-borne silt and is therefore active. Today, Sundarban bears mangrove forests over 4,264 km², which is more than 60% of the mangrove forests of India and hosts Sundarban Tiger Reserve, created in 1973 over 2,585 km² area (The only mangrove tiger reserve in the whole world). The remaining area of Sundarban comprises of the reclaimed inter-tidal zone supporting habitation and agriculture, yielding mostly a single crop of paddy.

Sundarban was once extended upto Calcutta. Many localities of Calcutta are named after tree types, e.g., Entally derived its name from Hetal tree, Goranhata (Gariahata) from Goran tree and Garia from Garia tree. This suggests that these areas were once covered with these trees from which they derived their name. The discovery of Sundari tree tissue from metro railway ditches proved that Calcutta was once (600-700 years ago) a part of Sundarban mangrove forests. Even now, a magnificent species of Sundari tree that once gave rise to the name of Sundarban can be found in the compound of the national library of Calcutta.

Sundarban has derived its name from Sundari trees (Heritiera littorails), a sturdy variety of evergreen tree with buttressed trunks (50 feet tall), spreading roots along the soil surface. It is now becoming so rare on the West Bengal side of Sundarban that it qualifies as an endangered species. Several other hypotheses have been proposed by other authors about the origin of the name. According to Pargiter (1889), Sundarban came from 'Samundraban' ('Samundra' means sea and 'Ban' means forest) which later corrupted into Sundarban. Alternately, it has also been suggested by some that since Royal Bengal tiger is nicknamed 'Sundar' (beautiful) the forest which is the kingdom of tiger is named as Sundarban (Mukherjee, 1983). Sundarban also means beautiful forest. There is no denying the fact that there is much in these forests that is beautiful. The forest is dense, the vegetation is lush-green, the river is eternal and full of life.

The surface topography of Sundarban is low, it is just 3 to 4 metres above sea level and is, therefore, inundated by tide water almost everyday. A high tide comes twice a day flooding the land with salty water. Hundreds of streams and creeks crisscrossed the area breaking it up into numerous islands. In Sundarban, there are 102 islands of which 54 islands are devoid of forest completely. These 54 islands are inhabited by people and are encircled by 3,500 kilometres of embankment or bundh, mostly made of mud to protect land from saline water. During the high tide the sea pumps water into the rivers raising their level about 16-17 feet. The pressure has steadily been cutting away slices from the embankments, playing havoc with life and property ever year and throughout the year. The brunt of the floods comes in September-October when the rivers are in full spate and cyclonic storms from the Bay of Bengal are very frequent. The flood water spells disaster for not only the crops but also the soils. The soil becomes extremely difficult to work due to unusual salt concentration, particularly in the dry season. In monsoon when the surface salts are considerably washed away and leached to the lower sub-soil, salt resistant varieties of paddy are grown. Apart from paddy, other crops grown in Sundarban are *paan* (betel leaf) jute, mustard and vegetables. Farming is of course seasonal and depends upon the weather.

The Sundarban is very rich in wood, shrimp and honey. The trees are not of great height (12 feet) but very dense. It is so dense that it restricts vision even upto five feet and this makes men easy prey to tigers. There are 40 varieties of trees and shrubs. Sundari, Goran, Hetal, Golpata, Bayen, Gewa are the principal trees. There are also plenty Neepa trees whose leaves are at present used for roofing. Unlike other regions, the menace of deforestation and pollution has become a part of life. The native people of the region are forced to plunder the forest for honey, wood, fish and other products in order to eke out a living. An ecological study recently done by Professor Chowdhury, identified at least 86 species of fish and 46 species of crustaceans that had either become extinct or endangered in the Sagar islands. Industrial effluents from Haldia Wreaks havoc on aquatic animals including microscopic planktons, which play a key role in the marine ecosystem. In other areas of Sundarban, similar processes are going on but due to lack of study these are unknown to us.

In Sundarban, around 5 million people are living in adverse condition created by salt affected soil and wildlife like tigers, crocodiles etc. As many as 85% people are dependent upon agriculture on reclaimed land which mostly bears a single crop of paddy. Besides agriculture, many people are found to take the risk of cyclone for fishing and others enter the forests to collect honey and fuelwood, braving the man-eating tigers, dacoits and other fierce animals like crocodiles. In all 50% agricultural labourers are landless; 44% of population belongs to scheduled castes and scheduled tribes. Most of the people live below the poverty line. Incidence of suicides are rampant and this is more among women that men. The communication is poor and most of the area is inaccessible. Though some of the points like Kakdwip, Namkhana, Raidighi, Sonakhali, Canning, etc., are linked with Calcutta by metal roads, the communication in this area dissected by the network of numerous streams and creeks, is mainly by boats and limited number of launches sailing through the water courses.

Sundarban was earlier known as Kalikaban and was inhabited by some non-Aryan races who were cultured to some extent and knew how to fight with boats and elephants. Greek travellers reported that the soldiers of Alexander during invasion of India (321-325 BC) were afraid to proceed further after hearing the courage and fighting capability of the people of Gangaridi (lower Bengal adjacent to the river Ganga). The Gangaridi was in existence even in Murya period and was ruled by elected representatives as in democracy. Most of the early evidences about the existence of these people were waved away due to great subsidence in lower Bengal during the middle of 6th century A.D.

In medieval period, small settlements mostly in the form of garden houses, check posts, worshipping sites and even dacoit hideouts developed throughout the region. Raja Pratapaditya and Isakhan (Bara Bhuniyas), who rebelled against the Mughal emperor Akbar, ruled over the region (in the late 16th and early 17th century) and built many naval outposts and garden houses in the heart of Sundarban. Remnants of these settlements were found in the midst of thick forest by settlers at much later date. During the first half of the 17th century, the Mughal and Portuguese pirates began their frequent visits to this land and the damage done by these pirates is in itself a chapter of history. Alibordi Khan who ruled Bengal (1740-56) fought many battles to restore normalcy. After the battle of Palassy (1757), the whole area came under the control of the East India Company.

The present history of settlement in Sundarban is not very old, as all the cultivated areas and villages of Sundarban were part of the forest till 1780. Therefore, this is the most recently settled region in the whole West Bengal. During the British rule, after the enactment of permanent settlement, large forest areas of Sundarban were leased out to zamindars (landlords) who in turn distributed these lands to subsidiary landholders (jotedars) or share-croppers (bargadars) for reclamation of land. The distribution of land at that time was massive and the settlers came mostly from the neighbouring Midnapore district. The first settlers came to the marginal areas (above high water level). But, with the increase in number of settlers, the clearings of forest was not restricted to only high water level but also extended to low water shorelines. The settlement of Sundarban is established by constructing embankment along with high water line. Before the construction of embankment, individual houses were boxed off after clearing the forest by throwing thick earthen wall. With the passage of time these earthen walls were worn down and used as village link roads during rainy seasons. This man-made embankment resists silt-laden water from spilling over the land on either sides of the river. The process caused the upliftment of the river-bed and consequent flood and infiltration of saline water into the farming land. The premature reclamation has heightened the flood effect and turned many reclaimed lands into permanent marshes. Subsequently, the ownership of land passed on to small landholders from big landholders. The small landholders did not have the capacity to maintain embankment. Some marshes turned into fish farms (bherhys) by rich landholders while others become the breeding grounds of mosquitoes. In Sundarban, out of the total 31,864 hectares of water bodies 20,000 hectares are used as fish farms.

In Sundarban, clearing of forest for cultivation and human habitation has proceeded relentlessly from the north. For example, Gosaba island at the beginning of the last century was a dense forest where tigers and crocodiles held sway when Daniel Hamilton first established his model estate. Now 100 years later, the mangrove forest has completely disappeared from there. At the beginning of the British period, Tilmen Henkel (1781) was granted 144 leases and permitted to reclaim these tidal mangrove forests into fertile agricultural land and fisheries. Since then, the reclamation of the Sundarban has started. It took place in different phases as shown in Map 1. The first phase (1770-80) was very extensive in the north and south-west, covering large parts of the police stations Haroa, Bhangar, Kulpi, Hasanabad, Baruipur, etc. These areas were then above the high tide water level and so bad effects of premature reclamation were absent there. The second phase of reclamation (1780-1873) was not very extensive due to some demarcation problems of land. Under this phase some areas of Hingalganj, Minakhan, Canning, Joynagar, Mathurapur and Sagar islands police stations were reclaimed. The third phase (1873-1939) of reclamation was very extensive particularly in the northern and south-western parts. The reclamation over premature land would cause a host of problems and even jeopardise the stay of man in many areas. In the fourth phase (1945-1971) only a few pockets of land were reclaimed for refugee population in various police stations of Sundarban, viz., Hingalganj, Gosaba, Basanti, Sagar, Patherpratima, Kultali, Namkhana, etc. After 1971, there has not been only official reclamation. In Sundarban, people of different areas came and settled. In south and western part people came largely from Midnapore district across the river Hooghly. The central and northern part are occupied by the Calcutta suburb people with a sprinkle of Midnapore people. In the north-eastern sector, both before and after the independence of India, excess population from Jessore and Khulna districts (now in Bangladesh) filled up the area. The illegal infiltration of population from Bangladesh is still continuing at many places which leads to further premature reclamation, at least marginally. In Sundarban, there are ancestors of some tribal communities like Santhal and 'Orans'. They migrated from Chhotanagpur plateau of Bihar, brought as hired labourers during the Mughal and British periods. The incoming population and their greed for land set in a process of premature reclamation deep into the formation areas and thus

Map 1
Sundarban Delta

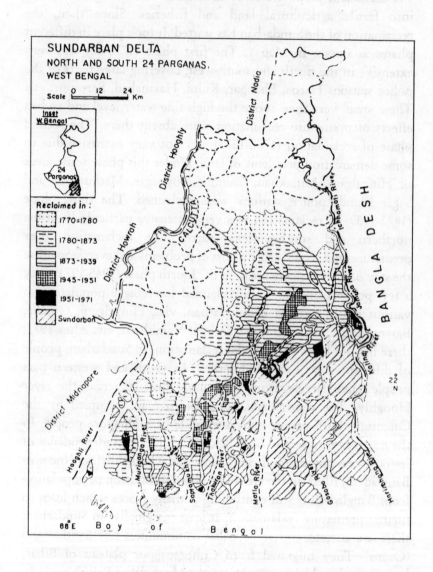

SUNDARBAN DELTA
NORTH AND SOUTH 24 PARGANAS,
WEST BENGAL

Scale 0   12   24
                              Km

Inset
W Bengal

24
Parganas

Reclaimed in :
1770–1780
1780–1873
1873–1939
1945–1951
1951–1971
Sundarban

has created a lot problems and is even jeopardising human habitat in many areas.

The problem of breaches in embankment in many places of the 3,500 kilometre long embankment is primarily caused by water currents. During high tides occurring at 12-hour intervals, a huge amount of water flows from the sea into the rivers causing the water level to go up by 16-20 feet above the islands. To avoid the consequent flooding and submergence of land, the early settlers built high embankments along the river. Pioneered by the then zamindars, the embankments were made of mud and bereft of any technology. But the ironic part is that breaches were as common then as they are now, more than two centuries later, since the embankments continue to be devoid of any scientific planning. The embankments are still made of mud in most cases and repairs, if at all undertaken by the State government, are superficial and patchy. Roughly, an approximate 80 paise is spent on the upkeep of each kilometre of embankment of which a major part goes for paying the staff and maintaining office. In view of this, the local people keep heaping mud to strengthen the embankments wherever cracks appear in them and have long given up expecting anything from the government. Breaches of embankment may happen at any time but the brunt of the floods comes in June-July when the monsoon sets in, and continues till November when the rivers are in full spate. The rains make the embankments weak and there is no telling when the river strikes. At places, bricks, sacks of sands, earth and bamboos and even wire meshes have been used to stem the erosion by the rivers. However, the efforts prove futile as they overlook one of the most basic causes of breach. In nearly all cases, the very foundation of the embankment is corroded by the strong currents at the base of the river. Thus, no matter what is heaped on them, the embankments still cave in. The wreckage of embankments is connected to underwater current and the patch repairing is useless since it cannot hold erosion. But, maintenance of embankment is most urgent for safety of settlements and agricultural fields in Sundarban region.

The failure of old embankment is a constant feature. The experts have expressed concern over the vulnerable condition of the existing embankments. The ring bundhs built to fortify villages at some blocks of Sundarban have not shown full proof to withstand the ravages of flood. In most places, the rivers have eaten into the ring bundhs, which in their precarious state remain only a feeble protection against the floods. Besides, construction of ring bundhs dislocates a section of the villagers who due to administrative apathy find no rehabilitation or even compensation.

There are some low-cost measures which can be adopted to stem the erosion of the embankments, albeit for the time being. For one, if mud has to be used to construct the embankment for lack of other expensive raw materials, let it be of the stronger kind and not the loose crumbly variety taken from the nearby cultivated fields or riverbanks as is the usual practice.

Secondly, the suggestion made by the foreign experts to construct new embankments behind the existing ones should be experimented. In October 1992, M. Zonneveld, an expert from Holland, visited many islands of Sundarban and had conversations with the local people and state irrigation authority. He recommended construction of retiring embankments, i.e., embankments at considerable distance behind the existing ones. The embankment should be built as far away from the main rivers as possible to minimise impact of the dashing waves. There should be a thick coverage of mangrove forest (Badaban) in the valley between the two embankments (See Figure 1). The proposal given by the foreign expert has not yet been implemented because the acquisition of land for the embankment and the intervening space for mangrove forest rests on the state government. The state government has not yet taken any initiative to acquire land for the construction of the embankment. The acquisition of land would lead to displacement of people, the rehabilitation of whom also creates a problem. Even then there is no guarantee (as expressed by experts) against breaches, for nature can be peculiarly obstinate. The sufferings of the people would remain as they were in the past. There are time to time piece-meal efforts on part of the state

Figure 1
Cross-section of the River

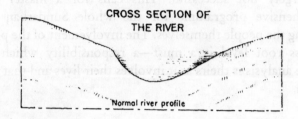

CROSS SECTION OF
THE RIVER

Normal river profile

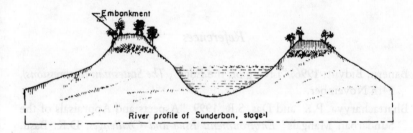

Embankment

River profile of Sunderban, stage-I

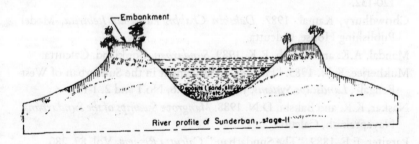

Embankment

Deposits (sand, silt, clay, etc.)

River profile of Sunderban, stage-II

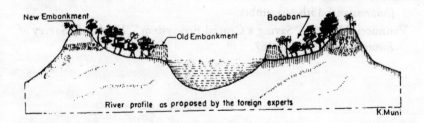

New Embankment

Old Embankment

Badaban

River profile as proposed by the foreign experts

K.Muni

government and the various non-government organisations (NGO's) for the construction of the embankment, but their efforts were largely not successful. This calls for a master plan, a comprehensive programme for the whole Sundarban region, involving the people themselves. The involvement of the people at the grass root level is a must—a responsibility which in the ultimate analysis is theirs as it involves their lives and that of their children.

References

Banerji, Bidyut. 1996. "Living on the Edge", *The Statesman (Impressions)*, 24th November.

Bhattacharyya, P.K. and Das, S.R. 1999. "Aspects and Appraisals of the Sundarban Mangals" *Environment: Issue and Challenges*, D.K. Basu, Amit Mallik and A.R. Ghosh (ed.), University of Burdwan, Burdwan. 120-132.

Chowdhury, Kamal. 1987. *Dakshin Chabbis Parganar Itebirtta*, Model Publishing House, Calcutta.

Mandal, A.K. and Ghosh, R.K. 1989. *Sundarban*, Bookland, Calcutta.

Mukherjee, K.N. 1983. "History of Settlement in the Sundarban of West Bengal", *Landscape Systems*, January, Vol.6, No.1 and 2. 1-19.

Naskar, K.R. and Bakshi, D.N. 1988. *Mangrove Swamps of the Sundarbans*, Nayaprakash, Calcutta.

Pargiter, F.E. 1889. "The Sundarban", *Calcutta Review*, Vol. 89. 280.

Sarkar, Sahini. 1998. "Shakled by Terror", *The Sunday Statesman (Impressions)*, 15th November.

Vannucei, M. 1991. "Saving a Crucial Eco-system". *The Hindu Survey of Environment*, Madras. 167.

17

Land Degradation Due to Indiscriminate 'Murrum' Extraction Near Durgapur Town, West Bengal

Susmita Ghosh and Subrata Ghosh

Land degradation is the wearing down of a land surface by some erosional process. When degradation is associated with desertification, it necessarily implies adverse changes in the environment, and the rate becomes alarming particularly when activated by man. The consequences of extortion of the land for man's immediate needs could be dangerous. It reduces the economic utility of the land, thereby making it unfit for any further use. The extortion of land by the thoughtless action of man reduces the area to a badland, carrying the spoils of the shortsighted economic activity, thus reducing the potential urban or agricultural growth zones. Ideally any economic activity should protect the potentiality of the land while utilising its wealth, either by restoring the landscape and stability of the land or by improving it through re-landscaping.

The study area is widely covered by laterites of varying composition. Laterite is extracted for different purposes, such as, making large building blocks (laterite is easily cut out from the ground and it hardens like a stone on exposure), alumina extraction (when bauxite occurs), collecting red and yellow pigments (ochres) from clayey and ochery portions, gathering colloidal constituents which are used for decolouring oils, extracting titanium, manganese, etc. (where concentration is high), gravel and pebbles (where they concentrate in distinct beds) and for 'murrum' (for paving unmetalled roads which are frequently found in rural India). In our study area the lateritic beds are being excavated mainly for 'murrum' and pebbles, without assessing other potentials.

The present study probes into the problems caused by the extraction of lateritic material, from areas in and around Durgapur town, in West Bengal, and suggests some remedial measures for checking further deterioration and ultimately for restoring the land. This problem was identified while carrying out an environmental mapping project for the Durgapur industrial area. Laterite though not considered agriculturally fertile, can support dense teak and mixed forests if the top layer is not disturbed. Ironically, it is some of the constituents of the laterite that make it vulnerable to reckless exploitation thus aggravating land degradation. In this process the land is deprived of its topsoil, mineral wealth and the natural landscape is destroyed. The changes in topography lead to disruption of drainage.

Study Area

Location

Durgapur town (recently accorded a municipal corporation status) in Barddhaman district of West Bengal is situated on the northern banks of river Damodar, 164 kms, north-west of Calcutta. The municipal area (13,825 acres), which holds most of the industries, was carved out of Durgapur-Faridpur block. Today the much smaller remnant of Durgapur-Faridpur block lies north of the municipal area whereas the Andal block is on the west and Kanksa block to the east. In a way, the study area may be considered as occurring in the interfluve area between Damodar and Ajay rivers.

Drainage and Physiography

The northwest corner of the study area is the headwater region of the Kunur Nala (tributary to river Ajay) and Tamla Nala (tributary to river Damodar), both seasonal streams which are loaded with sewage and industrial effluents. Immediately north and west of this area lie the numerous coal pits of Ranigunj coalfields. These coalfields are under the control of Eastern Coalfields Limited. The general elevation of the land ranges from 55 metres to 122 metres above sea level. The laterites are found mostly in the relatively higher areas and the excavation activity is at its peak in the highest (near 120 m) area, near a village called Pahari. For ease of understanding, henceforth the different parts of the study area shall be referred to according to the mouzas.

Climate

Geographically the area is most tropical, as the Tropic of Cancer (23°30'N) passes through the town. The sea coast being at least 220 kms away at the Bay of Bengal, somewhat extreme tropical climate is experienced here. The winter is cold and dry (mean lowest temperature is 4-5°C, with average 59 mm rainfall from Nov-Feb). The summer is very hot and dry (mean highest is 40-46°C, relative humidity 42-55% with average 187 mm rainfall and evaporation at the rate of 6-7.6 mm/day during March to early June), the dry spell is infrequently broken by some cyclonic thunder storms. The mean total rainfall for the monsoon period (June to September) is 985 mm when day temperatures come down to below 40°C and evaporation rates are close to 4 mm/day. Such climatic conditions have provided favourable condition for laterization

Geology

Geologically the area is significant as the alluvial sediments of the Bengal Basin are juxtaposed with the Older Gondwana deposits on the west. The western boundary of the Tertiary alluvial sediments runs across Durgapur (in almost N-S direction) while the eastern boundary of the Gondwana deposits lies under the veneer of sediments just east of Durgapur (near Panagarh railway station), imparting an interesting geology to the area. The Gondwana Basin Margin Fault Scarp Zone runs almost N-S about 20 kms east of Durgapur railway station. The Durgapur area is mainly underlaid by Quaternary, Tertiary and Gondwana formations (Table 1).

Table 1
Geology of the Area

Age	Formation	Lithology
Holocene	Panskura	Clay alternating with silt and sand and detrital laterite (near the river valleys)
Pleistocene	Sijua	Clay with caliche concretion (near the banks of river Ajay)
Cainozoic	Laterite	Primary laterite (covering large part of the area)
Jurassic	Durgapur bed	Coarse feldspathic sandstone (west of Durgapur)
Carboniferous to Triassic/ Permian	Panchet	Red shale and sandstone (west of Durgapur)
	Ranigunj	Fine grained sandstone, siltstone (around Ranigunj, west & north of Durgapur)

Holocene (younger alluvium) and Pleistocene (older alluvium and detrital laterite) formations represent the Quaternary formations. The younger alluvium occurs in the valley fills, riverbeds, banks, and lower flood plains. These are mainly fine to medium coarse sand, silt and clay. The Gondwana group consists of lacustrine or fluvial deposits. They generally become finer upward, with intercalated plant remains which form the coal beds (Krishnan, 1982). Within this group Damuda system (Table 2), deriving its name from the river Damodar, is most extensive in thickness (upto 2,000 metres) and it is best developed.

The Damuda sediments reflect warm and humid conditions of deposition during Permian times that permitted the growth of luxuriant vegetation, and marked the peak of *Glossopteris* flora. The Barakar series contain coal seams; it is followed by the Barren Measures, which are remarkable by the absence of coal and consisting mainly of ironstone shale. The coarsening upward of the ironstone shale mean that they were deposited in a quiet, reducing, lacustrine type of environment, generated by gradual progradation of a delta into a shallow marine or deep-water lake (Sengupta and Mondal, 1987). The Ranigunj system overlies the Ironstone shale,

Table 2

Gondwana Succession in India (after Krishnan, 1982; * R. Kumar, 1985)

Age		Formation	Example from the study area
Cretaceous	L	Umla	
		Jabalpur	
		Chugan	
Jurassic	U	Rajmahal	
	M	Hiatus (?)	
	L	Mahadeva	
		Supra Panchet	
Triassic	Rh	Panchet	Fluvitile origin
	U		
	M		
Permian	U	Ranigunj	
(Damuda	M	Barren Measures	Ironstone Shale, Kulti
System)	L	Barakar, Kaharbari	
Carboniferous	U	Talchir, Talchir-tillite	

which are similar to Barakar series in character. The Panchet series (of Triassic age), which overlaps the Damuda system (mostly the Ranigunj system and sometimes the Barakar series), reflects a drier climate as they enclose almost no carbonaceous matter. Panchet series marks the end of the *Glossopteris* flora and the emergence of the *Ptilophyllum* flora. The Supra-Panchet deposits (otherwise known as Mahadeva series) are sediments of an arid climate. They consist of ferruginous and feldspathic sandstone, such as Durgapur sandstone beds. Around Durgapur area we find the Durgapur beds outcrops in the city centre and along the G.T. road. The Ranigunj and Panchet series are found on the northwestern part. In between the Durgapur beds and the Panchet beds a vast area is covered by laterite. Some of this is of detrital type in the valley fills. There are some prominent pebbly layers within this laterite, particularly in Kamalpur and Hetedoba mouzas, just north of the Durgapur Steel township. This pebbly layer could be derived from a *conglomerate* bed, which may be correlated to the Bagra conglomerate (an upper formation of the Mahadeva group). Today, this supposedly conglomerate bed is deeply weathered to form a pebbly layer within the laterite. However, in some cases it could be fluvitile material derived from the Rajmahal Traps.

Laterites

Laterite is the residual weathered material formed under special climatic conditions in tropical regions. Laterite consists essentially of hydrated iron oxides, bauxite of hydrated aluminium oxides, the most common impurity in both being silica; aluminous laterites and ferruginous are quite common. Laterite is a porous, pitted, clay-like rock with red, yellow, brown, grey and mottled colours depending in some measure on the composition. It has a hard protective limonitic crust on the exposed surface that is rough. As mentioned earlier, much of the study area is covered with laterites. Laterite may be developed over a large variety of rocks (alkali rocks, intermediate and basic igneous rocks, dolerite/basalt, gneissic rocks rich in feldspars, sedimentary rocks including shales, sandstones and impure limestones). The laterite capping may be as thick as 50 to 60 metres. There is usually a layer of highly ferruginous material at the surface, below which there is a bed of aluminous laterite or *bauxite*. These grade further below into a *lithomargic clay* layer, which gradually merges into *saprolite* (deeply weathered rock that retains the outward structural form of the original rock commonly found in tropical climate), and finally to the unaltered parent rock. Laterite also occurs in the plains and at the base of hills, these being of secondary origin, derived from the high level laterite and re-cemented after deposition in the valleys or plains. Low level laterite is therefore mainly of detrital origin. Detrital laterite is heterogeneous, whereas primary laterite is compact and fairly uniform in composition (Krishnan, 1982). Occasionally when bauxite and laterite are eroded and redeposited it is unlikely that they have been transported over great distances.

They become mixed with other sedimentary material, to form such rocks as bauxite clays. Spheroidal masses sometimes develop in bauxites and laterites, and the term *pisolitic* has been applied to these rocks, although there is no evidence that the structures have been produced in the same way as true pisolites. These have been formed during Upper Tertiary or later; probably majority of it during Pleistocene and the process is still active.

The essential climatic feature for lateritisation is a well-marked division of the year into wet and dry seasons. During the wet season, leaching of the rock occurs and during the dry season the solution containing the leached ions is drawn towards the surface by capillary action, where it evaporates, leaving salts to be washed away during the next wet season. Thus the whole zone, from the lowest level to which the water-table falls to the highest level it reaches, is progressively depleted of the more easily leached elements, e.g., sodium, potassium, calcium and magnesium. A solution containing these ions may have the correct pH to dissolve silica in preference to aluminium oxides or iron oxides. Thus the residuum consists mainly of these two oxides. Which of them is predominant depends mainly on the characters of the original rock type, e.g., basic igneous rocks and other iron-rich types tend to yield laterite, while granitic rocks and others low in iron yield bauxite. The former is true in our study area. An essential property of the rock is that it should maintain a porous structure during leaching process, so that the fluids may circulate freely. For this reason hard rock exposures of a particular rock type lie juxtaposed with laterite, as is found in our study area.

Lateritic soil (or *latosol*) is a zonal soil formed in lowland regions of humid tropics over laterites. Such soil is rich in iron and aluminium with some titanium and manganese. An important point is that latosols often support dense forests but once these forests have been removed the reduction of shade and humus accelerates lateritisation and the soil soon loses fertility and friability, and becomes agriculturally poor.

Laterite capping formed from various types of rocks and also of detrital nature are found across mouzas of Faridpur, Mahishkapur, Parulia, Sovapur, Parangunj, Chakgopaldebpur, Bijupara, Kamalpur, Parulia, Bansol, Hetedoba, Bansgarha, Shyampur, Bansia, Jamgarha and Tilabani in the study area. The area east of Jemua has latosols and alluvium. Clay beds exist north of Jemua, close to Kunur Nala. The Supra-Panchet outcrops of Durgapur are lateritised deeply and *saprolite is* observed (e.g., exposure in a 'murrum' quarrying pit along Sadhu Ram Chand Murmu road). The pebbly layers in the laterites in this area may suggest a detrital type, whereas composite primary laterite, with limonitic crust, is found further north in Jamgarha mouza.

Scenario of the Affected Areas

In the heart of Durgapur town, 'murrum' is collected from large pits using machines. This land is owned by SAIL and will be handed over to ADDA for development into residential plots. The pits now existing will be filled up with wastes veneered with some soil. So, hopefully there will not be much problem. However, in Kamalpur, Hetedoba, Chapabandi, Bansgarha, Bansia and Parulia the scenario is different. Pits are randomly dug out and gravel (including rounded pebbles of varying sizes and origin) are extracted. It is reported that these pebbles are exported for industrial use. The remaining laterite is generally left dumped as land spoils from the quarrying activity. The land is being raped here by the reckless economic exploitation. The area is pockmarked with pits and devoid of vegetation, turning it into a kind of 'moonscape', like badlands. Badland topography illustrates one set of conditions, which lead to fine drainage texture. Impermeable clays and shales, sparse vegetation and prevalence of dash rainfall have been responsible for extremely fine drainage texture (Thornbury, 1969). These areas necessarily have very high soil erosion rates. The place is so dangerously pitted that it is not possible even to walk around easily.

Table 3
Geographical Area of the Affected Mouzas

Name of mouza	Area (hectares)	Affected (% of total mouza area)
Kamalpur	784.49	75
Hetedoba	368.72	75
Chapabandi	134.64	60
Bansgarha	485.30	55
Bansia	299.31	90
Parulia	970.79	60

The propriety and history of extraction activity explains the recklessness with which it is carried out. The area was given on a long-term lease to the operators in 1911 by the then Raja of Barddhaman. Even after the abolishment of zamindari system, independence and finally the transfer of this land to SAIL, forest department and the gram panchayat, the right of the leaseholders prevailed. There was an attempt to stop the activity but a petition by the sharers of the operation resulted in High Court order allowing the quarrying operation to continue. Indeed some thousand families are dependent on this operation and there is no other economic activity in the area to turn to. There is also political interference to promote the activity. There is continuing conflict between the forest department and the operators as the activity adversely affects the plantations in the area. Even when some mutual understanding is arrived at between the two conflicting parties it falls through, as there are so many shareholders with none to shoulder the responsibility.

Mapping Procedure

As mentioned earlier, this problem area was identified while carrying out an environmental mapping project for Durgapur industrial area. Hence the mapping started with cadastral mouza maps (16 inches to 1 mile scale, i.e., 1:3,960) followed by fitting its details on to police station (P.S.) maps (4 inches to a mile, i.e., 1:15,840) as the outline development plan maps were prepared at this scale. Topographical maps, at 1:50,000 scale, were also consulted but much additional details are not available from them, apart from contours and geographic coordinates. IRS 1B and 1C imagery of three different seasons were visually interpreted and classified for distinguishing bare lateritic areas, types of vegetated areas, inhabited areas, water bodies, etc. Apart from conventional ground control points at road, rail or canal intersections, ponds and tanks served as very useful control points as well as reference points during field survey. The graticule of latitude and longitude were drawn at 2°30' interval. Classified satellite imagery (derived from IRS of May 1994, Aug 1996 and Feb 2000) was enlarged and superimposed on the P.S. maps. Detailed field data were drafted on to mouza maps and then incorporated into the P.S. maps (at 4 inches to a mile). In this way field checking and field data could be accurately placed on the map. Such a mapping procedure would allow the various data layers to be automated and analysed in a geographical information system (GIS). Further update of data would indicate changes in the landscape, vegetation cover, excavated area, water bodies, etc.

Consequences

The adverse effects of the 'murrum' or gravel/pebble extraction are:

(1) Rendering an ugly desolate badland type landscape to the area.

(2) Making the area unfit for anything (such as agriculture, plantation, construction, recreation, etc.).

(3) Deforestation: Scraping the land of all vegetation, harming the fringes of adjacent plantations and thereby affecting the ecology.

(4) Desiccation: Although the pits do allow some water to be stored, it deprives the remaining higher area from groundwater and soil stability. This aggravates land degradation and causes trees to fall.

(5) Since this area (around Pahari and Hetedoba) has highest elevation in the area within 10 kms (northward) and 15 km (southward) radius, it is the source region of Tamla and Kunur 'nalas'. Degrading of the land here adds to the sediment load of the streams, simultaneously it reduces their energy of flow (by

decreasing relative relief and increasing load). It also truncates the headwater catchment zone.

(6) The deep pits increase the chances of groundwater contamination. We must remember that it is not at all easy to decontaminate a particular groundwater reservoir.

(7) The topsoil is removed and destroyed, which would take centuries to develop again. This problem is further discussed in connection with lateritic soils.

(8) The unwanted lateritic matter is dumped here and there, but there may be more important mineral wealth in them.

Reclamation of the Affected Land

Although the lateritic area is badly affected, all is not yet lost as the land can be recovered by some remedial measures as mentioned below. Apart from technical measures, the implementation aspect must be remedied for the remedial measures to be effective.

(1) The excavation operation should be controlled by the gram panchayat and directed towards the well-being of the villages.

(2) As with 'polluter pays' policy, the damaged land should be restored by the people undertaking the excavation. It is necessary to level the land as much as possible.

(3) The excavation should proceed only in a particular direction so that adjacent abandoned pits can be restored (levelled) to useful land in a planned manner.

(4) Afforestation and protection of existing plantation is necessary in these areas. Social forestry would do well as this area is suitable for trees (such as jackfruit, mango, jamun/blackberry, neem, sajna, bel, kul/plum, figs, sal and segoon/teak).

(5) In some cases near the peripheral zone, the gravel bed could be removed so as to expose the underlying fine silty material that can be used for agriculture. At present such fine sediment is exposed in the intervening valleys whose narrow floodplains are cultivated.

(6) Some pits in consolidated lateritic beds at lower elevations could be turned into storage tanks that would hold back the

water of the monsoon rains to be used in the dry season. This could help in groundwater recharging. Trees could be planted successfully on the edges of the tanks.

(7) Some pits can be used as landfills for non-hazardous municipal or industrial wastes or mining overburden. Thereby the area could be levelled and given a soil veneer. Such land can be used for plantation, recreation or construction.

(8) For future excavation activity it is suggested that the displaced topsoil (with humus and vegetation) be kept aside and used for surfacing the filled up (restored) pits. It would have been excellent if the same soil could be used for veneering the area from which it was collected. But, since the process of excavation and gravel sorting is a long one it may not be possible to conserve the nature of the displaced topsoil for so long. Additional organic matter may be added to the veneer so as to promote future soil development and vegetative growth.

(9) Fencing of the damaged land is necessary, especially during restoration process, as much as it is necessary to fence the plantation areas for their protection.

Conclusion

Land degradation is taking place at an accelerated pace in parts of this study area due to excavation of lateritic land for 'murrum' and gravel (pebbles) used for road paving, construction and industrial purposes. The original vegetation has been removed and the young plantations are also being adversely affected by land erosion. As a result a kind of desertification is taking place. Moreover, the excavated area wears an ugly desolate badland-like look with pits of varying depths and sizes densely covering the area. This has rendered the landuseless until some remedial measures are taken.

There are a variety of ways in which these gravel and laterite ('murrum') pits can be dispensed with or remedied. These are listed in the previous section. Important among them are planned and unidirectional excavation rather than randomly, remedial action by the user, land restoration by landfill or levelling down (as applicable), promotion of vegetative cover, used as tanks to store irrigation water, veneering by topsoil, etc. Thus, the economic viability of the land would be enhanced. These lands, lying at the periphery of Durgapur, are most likely to be in great demand for various productive landuses such as industrial, residential, agricultural or forestry, if reclaimed or developed.

It is hoped that the concerned authorities would take necessary actions to prevent this destruction of land.

Acknowledgement

We thank Dr Arunaditya Majumdar (Principal, Durgapur Govt. College), Dr. B.C. Poddar (Ex-DDG, G.S.I.) and Durgapur Block Land Records Officer for their kind suggestions while writing this paper. We are also thankful to the Department of Geology of Durgapur Govt. College, for providing the facilities to draft this paper.

References

Krishnan, M.S. (1982). *Geology of India and Burma*. C.B.S. Publishers and Distributors, New Delhi.

Kumar, R. (1985). *Fundamentals of Historical Geology and Statigraphy of India*. Wiley New Delhi.

Ray, S. 1979. *'Bhlltattiker Chokhe Paschim Bangla'* (in Bengali), West Bengal State Book Board.

Sengupta, S. and Mandal, A.K. (1987). Cyclicity as clue to depositional environment: A study from the Ranigunj coalfield. In 'Geological Evolution of Peninsular India — Petrological and Structural Aspects' (Prof. Saurindranath Sen, 66th Birthday Commemorative Volume) AK. Saha (ed.) Recent Researches in Geology, India. Volume 13, pp 143-149.

Smith, J. (ed.) (1984). *Dictionary of Geography*. Arnold Association, New Delhi.

Thornbury, W.D. (1969). *Principles of Geomorphology*, Wiley, New Delhi.

Whitten, D.G.A. and Brooks, J.R.V. (1979). *A Dictionary of Geology*. Penguin Books Ltd., U.K.

18

Assessment of Soil Erosion and Degraded Land: A Case Study of Dumka Subdivision, Jharkhand

V.C. Jha and Saswati Kapat

Soil erosion is an ever present phenomenon and its continuous process, whatever may be its rate, brings about degradation of land. At present the accelerated soil erosion vis-a-vis land degradation, brought about by natural morphodynamic process coupled with man's excessive land abuse, induces gradual conversion of usable land into unusable one or productive land into less productive or unproductive one. Such adverse process of transformation of land may be viewed as one form of 'desertification' (Siva Kumar et. al., Nagarajana et. al., 2000, p. 7-8).

In India, about 130 million hectare of land (45% of the total area) is affected by soil erosion (Environmental statistical compendium, 1998). It is corroborated by the findings of several researchers (Bali and Kanwar, 1977, Sehgal and Abrol, 1994, Mandal et. al., 1996). They are of opinion that water erosion induced soil degradation is the major problem of India, particularly in the sub-humid and semi-arid areas.

In this context, an understanding between the erosion process under different conditions of the nature of soil, its usability and the management techniques, is required for the prevention and reduction of the soil loss rate to acceptable amount. It is important to explore how well or in what form, processes are understood and what the conditioning factors are affecting that understanding (Perez, 1998, p. 294).

All these considerations should be taken into account for the optimum use of land to meet the growing demand of the need of ever increasing population of our country.

From the perspective of the above views, this paper attempts to fulfil the following aims:

- The estimation of the amount of soil loss.
- The classification of the study area on the basis of the estimated soil loss.
- The identification of the productive/usable areas that are likely to be affected and the areas that have been moderately or severely damaged by the process of soil erosion/soil degradation in relation to climatic, geomorohic, pedogenic, vegetative and anthropogenic attributes.

Materials and Method

The expected soil loss is determined using universal soil loss equation (USLE) evolved by Wischmeier and Smith (1958 and 1978).

$$A = RKLSCP$$

A = Predicted annual average soil loss (metric tonnes/ha)

R = Erosivity (rainfall and runoff) factor

K = Soil erodibility factor

L = Slope length

S = Slope gradient/steepness

C = Cover and management/cropping management factor

P = Erosion control practice/conservation practice factor

The value of 'A' has been found out on the basis of the following estimation:

- Rainfall and runoff erosivity index (R) derived on the basis of rainfall (intensity for 30 minutes period and kinetic energy) data for ten years (1990-1999) from IMD and Dumka district-weather station and runoff data from the soil conservation office, Dumka.

- Soil erodibility factor (K) obtained from Wischmeier and Smith's nomograph/table.
- Topographic factor LS (Slope steepness 'S' and length of slope 'L') derived from soil profile report from agricultural office, map and toposheets on scale of 1:25,000 and 1:50,000 and nomograph of Wischmeier and Smith.
- Crop management factor 'C' and conservation practice factor 'P' obtained from landuse data/map (district office, Dumka) and report on soil and cropping system (district office, FAO, 1992 and Wischmeier and Smith's table).

Twenty plots of approximately one hectare and each grid point have been chosen for the study of the soil loss. Moreover, plot-wise information of related factors has been superimposed on the toposheets (on the scale of 1:25,000 and 1:50,000, number-series of 72 P). Plots have been selected on contrasting terrains (Table 1).

Table 1
Plot Characteristics of Dumka Subdivision

Block	Terrain Characteristics			
	Location	Plot no.	Nature of topography	Landuse
Dumka	Hijla	1	Level slope near the Mayurakshi river bank	Cropped land
	Bhurkhanda	2	Undulaing land affected by gully erosion	Waste land
Gopikandar	Dumertala	1	Foot hill zone of the south-west Rajmahal highland	Forest cover
	Khoirasol	2	Steep slope	Forest cover
Jama	Silandah	1	Undulating terrain affected by rill erosion	Arable land (double cropping)
	Bahiar	2	Undulating terrain	Fallow land
Jarmundi	Lagwa	1	Undulating terrain	Cultivated land (tripple cropping)
	Mahua	2	Undulating terrain	Current fallow

Contd...

Contd...

Kathikund	Aganbane	1	Hill range slope	Thick forest cover (with considerable trees)
	Dudhia	2	Foot hill (gullied)	Degraded forest cover
Masaliya	Ranga	1	Gentle undulaing terrain	Cultivated land
	Mahespatha	2	Foot hill, rock debris in the vicinity of plot	Cultivated land
Ramgarh	Silpahar	1	Undulating terrain	Cultivated land (tripple cropping)
	Naya Chak	2	Undulating terrain	Fallow land
Raniswar	Ektala	1	Level surface near the Mor river bank	Arable land (single cropped)
	Gobindpur	2	Level sloped terrain	Fallow land
Saraiyahat	Hansdiha	1	Undulating terrain	Barren (nearly)
	Kallpur	2	Undulating terrain	Slightly gully affected arable land
Shikaripara	Naupahar	1	Undulating terrain	Continuous fallow land
	Thilidabar	2	Undulating terrain	Double cropped land

Description of Study Area

- Dumka subdivision (district Dumka, Jharkhand, India) belonging to the Rajmahal highland lies between 23°59′ N to 24°40′N latitude and 86°65′E and 87°42′ E longitude, with an area of 302,271.35 hectare.
- The area is predominantly of gentle undulating terrain having four prominent residual hill ranges and a number of scattered hills. The altitude of the area ranges between 502m (Mahuagarhi) to 100m (Mayurakshi river bank)
- The majority of the area is composed of gneissic rock with some occurrence of gabbro, dolerite, sandstone, shale, etc.
- It is representative of dry sub-humid monsoon climate (according to moisture index-13.28 under Thornthwaite formula) with monthly average temperature ranging from 15.17°C (January) to 30.15°C (May) and monthly average rainfall from 2.02 mm (December) to 359.2 mm (August); average rainfall intensity from 1.8 mm (December) to 19.9 (August); potential monthly average evapotranspiration between 23mm (January) and 182mm (May).

- Considerable runoff is generated only in rainy season with the amount of rainfall ranging between 153.75mm/day (August) and 17.13mm/day (October)
- The soilscape is constituted mostly by the red-yellow and lateritic soil (Venkataraman and Krishnan, 1992, p. 479) under the soil series of Dumka, Pusaro, Lachhimpur, Karaya, Sarua, Hathipathar, Kanchanpur, Balia, Sirsia, Haripur, etc. Soils vary in colours from yellowish/dark-brown to grey-brown. Soil texture of surface and sub-surface vary from gravelly loam, sandy-clay loam (dominant texture in surface horizon) to clay loam (dominant texture of sub-surface horizon). Granular and dominant moderate sub-blocky structure in surface horizon and blocky in sub-surface horizon are the characteristics of the existing soilscape. The B horizon is enriched with sesquioxides. The rate of infiltration is between 7mm/h to 12mm/h.
- It is mostly drained by the Mayurakshi and its tributaries.
- The average slope varies between 1°5' and 15°14'.
- Vegetation is sparse/scanty except on some hill slopes which bear forest cover. It is mostly deciduous and shrubby in character.
- Single, double and triple cropping systems are practised.
- The classes of landuse of the area are as follows: net cultivated area (41.56%), current fallows (20.04%), area under forest (10.56%), area not available for cultivation (15.36%), uncultivated land (9.12%), and area under wasteland (3.36%) (Report of Lead Bank, Dumka, 1998).

Result and Discussion

The difference in rainfall, runoff, micro-topography, slope/gradient, soil characteristics, surface cover, etc., have brought about differences in the values of individual factors of USLE in 20 sample plots and also additional grid points examined.

1. Rainfall-runoff erosivity factor (R)

The index factor 'R' measures the erosion force of rainfall and runoff. But, much emphasis is given on rainfall aspect for measuring R factor (Schwab, 1901, p. 103).

The average R value of the selected plots in different blocks ranges between 104 (Jarmundi) and 210 (Raniswar) (Table 2). It lies around 140 in most of the blocks except Kathikund, Shikaripara and Raniswar which bear 201, 205 and 210 respectively. The distribution of R factor exhibits its gradual decrease towards the west from the east coinciding with decreasing annual average rainfall (1120.34mm to 803mm; from east to west).

2. Soil erodibility factor (K)

K factor indicates the inherent erodibility of soil.

The average value of K factor of the plots in all blocks varies from 0.14 (Gopikandar) to 0.48 (Ramgarh). This range refers to readily infiltrated soils (K <0.2); soils with intermediate infiltration capacities (K -0.2 to 0.3); and more easily eroded soils with low infiltration capacities (K >0.3) as suggested by Brady (1990, p. 439). Relatively high K value (>0.3) of the plots in blocks of Jarmundi, Saraiyahat and Ramgarh indicates higher erodibility of soil owing to its textural character (sandy-clay loam), scanty vegetation cover, presence of nearly barren and also fallow land which in turn make the area very susceptible to gully/sheet/rill erosion. The remaining blocks except Gopikandar and Kathikund restrict their K value to slightly >0.2 in view of their characteristic soil in particular. The blocks of Gopikandar and Kathikund bear low K values which are 0.14 and 0.15 respectively, due to their location on hill slope having forest cover and soils with gravelly and sandy-clay loam texture which help in rapid infiltration inspite of having heavy rainfall.

3. Topographic factor (LS)

The slope length - slope steepness value is the ratio of soil loss from the slope in question to the soil loss from the reference. The reference slope has an LS factor of 1 (Shickluna et al., 1983, p. 430).

It ranges from 0.34 (Raniswar) to 1.05 (Kathikund). The selected plots in the blocks of Dumka, Masaliya, Gopikandar and Kathikund bear relatively higher topographic factors which are 0.72, 0.76, 1.03 and 1.05 respectively, as they belong to Rajmahal hill range and also hill ranges in proximity to Masanjor. The plots in the remaining blocks bear LS value ranging between 0.34 to 0.69. In fact, rolling topography produces such a range of topographic factor.

4. Vegetative cover and management factor (C)

C factor is the ratio of soil loss from areas with protective cover to the corresponding loss from areas without it (Shickluna et al., 1983, p. 432).

The obtained values of C factor of the selected block-wise plots range between 0.17 (Kathikund) and 0.74 (Ramgarh). The sample plots in the blocks of Kathikund, Gopikandar and Masaliya have low C value (< 0.2) because of their locations on the forested slope or in the vicinity of forest area. Plots in Dumka, Jama, Shikaripara, Raniswar blocks bear moderate C factor ranging between 0.26 to 0.50. Ramgarh, Jarmundi, and Saraiyahat blocks produce higher C values varying from 0.57 to 0.74 in view of fallowness or continuous ploughing of plots.

5. Conservation practice factor (P)

It reflects ratio of soil loss with a given support practice to the corresponding loss without support practice (Brady, 1990, p. 443).

P factor varies from 0.5 to 0.6. This range of P values has been derived from all the sample plots taken from each block of Dumka subdivision.

Table 2
Ranges of Average Values of Factors Related to Soil Loss Equation of Dumka Subdivision

R	K	LS	C	P
104 - 210	0.14 - 0.48	0.34 - 1.05	0.17 - 0.74	0.5 - 0.6

6. Universal soil loss in tons/hectare (A) derived as an annual average

The value of A ranges from 8.41 tons/ha (Kathikund) to 29.03 tons/ha (Ramgarh) (Table 3).

Table 3
Annual Average Soil Loss (A)

	Blocks	A values (tons/hectare)	Classification of A values	Remarks
1	Kathikund	9.41	< 10 tons/ha - least/low soil loss	Low soil loss
2	Gopikandar	8.41		Low soil loss
3	Masaliya	14.03	10-15 tons/ha - moderately low soil loss	Moderately low soil loss
4	Jama	18.10	15-20 tons/ha - moderate soil loss	Moderate soil loss
5	Dúmka	19.45		Moderate soil loss
6	Raniswar	18.75		Moderate soil loss
7	Shikaripara	22.50	20-25 tons/ha - moderately high soil loss	Moderately high soil loss
8	Saraiyahat	26.01	> 25 tons/ha- high soil loss	High soil loss
9	Jarmundi	27.31		High soil loss
10	Ramgarh	29.03		High soil loss

Kathikund and Gopikandar produce small amount of soil loss, estimated to 9.41 tonns/ha and 8.41 tons/ha respectively, inspite of having high rainfall. It is the forest cover and sandy, gravelly and sandy-gravelly loam soils borne by these two blocks that allow considerable infiltration of rainwater. Holes on the side wall of the gullies in vicinity of sample plots suggest this fact. Plots in the block of Masaliya are characterised by moderately low amount of soil loss (14.03 tons/ha/year). The plots in the blocks of Dumka, Jama, and Raniswar display moderate amount of soil loss (18.10 tons/ha/year to 19.45 tons/ha/year). The block of Shikaripara has moderately high value of A (22.5 tons/ha/year). The other blocks including Ramgarh, Jarmundi, Saraiyahat, produce high value of A ranging between 27.13 tons/ha/year and 29.03 tons/ha/year. It is interesting to note that cultivable plots lying within these three blocks with high value of A cause more soil loss liable to ill agricultural practices. It is obvious that maximum soil loss occurs during south-west monsoon periods of high rainfall.

Figure 1 exhibits that the area possesses four categories of soil loss (A), derived from sample plot investigation and also official

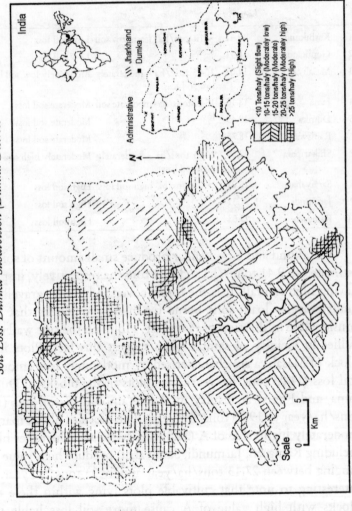

Figure 1
Soil Loss: Dumka Subdivision (Dumka District

data of sites other than sample plots of the study area. From this map, it is clear that the jungle blocks of Kathikund and Gopikandar and south-eastern part of Dumka (10% of study area) contain considerably low soil loss in amount; 30%, 35%, 10% of the area belong to the zone of moderately low, moderate and moderately high soil loss; and 15% of the total area is suffering from relatively high soil loss. High soil loss is widely noticed in the gully/rill affected area even in the arable land.

Moreover, annual agricultural production of the sample plots, collected during the period from 1990-1999, reveals its gradual decline in every year. It is obviously due to the continuous process of soil loss. If the soil loss continues at the estimated rate, it will bring this area to such a high rate of land degradation that the land may become totally unproductive or barren one assuming the nature of desert in near future.

From the above discussion, the following findings can be drawn:

- Forested slopes of jungle blocks produce least soil loss.
- Cultivable land beside barren or wasteland does not get rid of soil loss.
- Most of the sample plots are susceptible to moderate soil loss.
- High amount of soil loss is noticed in the blocks which have considerable cultivable land and wasteland.

Hence, possible ways should be explored and immediately implemented at the village level to mitigate the adverse effects of soil loss on land and its agricultural production system.

In this context, following measures can be taken into consideration:

- Proper bunding of the plots to minimise soil loss.
- Intensification of intercropping system.
- Introduction of agro-horticultural practices for increasing cover as well as income of the farmers.
- Raising of agro-forestry on the degraded land.

Acknowledgement

I gratefully acknowledge Prof. V.C. Jha and Shri G. Kapat (Retd.), Department of Geography, Visva-Bharati, for continuous encouragement and their critical comments.

References

Bali., Y.P. and J.S. Kanwar (1977): Soil degradation in India, in *FAO Soil Bulletin*, No. 34.

Biswas, et al. (1999): Prioritisation of sub-watersheds based on morphometric analysis of drainage basin, *Photonirvachak*, Vol. 27, No. 3, p. 155.

Brady, N.C. (1990): *The Nature and Properties of Soils*, Macmillan Publishing Company, pp. 438-445.

Nagarajan, M. (2000): Leisa India Supplement, March: Role of agroforestry in controlling desertification with special reference to Linn . . . in arid farming of West Rajasthan, p. 8.

Perez, J.D.C. Geo Okodynamic XIX - 1998: The use of pair-wise photographs for soil and water conservation planning: A case study of Guadalazara, p. 293.

Report of Lead Bank and District Statistical Department, 1991-1998.

Report of runoff, sediment yield landuse and soil, Birsha Agri-University.

Siva Kumar, et al. (2000): Leisa India Supplement: Landuse planning for the lands of the north western zone of Tamil Nadu.

Schwab, G.O. (1981): *Soil and Water Conservation Engineering*, John Wiley & Sons, pp. 100-109.

Shickluna R.W.M. et al. (1983): *Soils: An Introduction to Soils and Plant Growth*, pp. 428-433.

FAO (1992): Agro-ecological land resource assessment, World Soil Resource, Report No. 71/20; soil erosion and productivity, p. 46, FAO Roam.

Wischmeier, W.H. and D.D. Smith (1978): Predicting rainfall erosion losses. United States Department of Agriculture Hand Book, No. 538.

19

Land Degradation in Tropical Lands: A Case Study

V.C. Jha and K. Gupta

Land degradation relates to the loss of utility or potential utility or the reduction, loss or change of features or organisms, which can not be replaced.

Since last two decades it has been noted that every year millions of hectares of land suffer from physical, chemical and biological degradation throughout the world. This reduces their current and potential capability to support foodgrains, goods and services through agricultural engineering and other important usages. It also largely relates to human induced and natural phenomena in their geomorphic processes. The land degradation process has direct implications on socio-cultural components of the terrain.

Increased food production in tropical and sub-tropical countries is essential to feed the rapidly growing population. Intensification of agriculture, industrialisation and commercialisation have resulted in drastic changes in the land economics, directly affecting denudational processes. In recent years, land degradation has become a common and widespread problem due to climatic changes and anthropogenic interference through landuse, particularly in tropical lands.

Keeping in mind the above facts, the present study has been made. This paper deals with the factors responsible for land degradation in Dwarkeswar Basin in West Bengal, India and aims to assess the impacts of land degradation in society.

The present study proposes conservation management and comprehensive organisational inputs at various levels that will check the degradational process and in turn increase the sustainable benefits to communities, human and otherwise.

The study of land degradation, desertification, on one hand, and on the other the measures of conservation, rehabilitation need various types of spatio-temporal data in order to make sound environmental assessment and long-term projections. The use of remote sensing and GIS is indispensable in this process. This paper is based on primary field survey information and auxiliary secondary supportive database through remote sensing, GIS, statistical and cartographic techniques.

Study Area

The Dwarkeswar Basin (22°45′ N to 23°35′N and 86°30′ E to 87°45′ E) covers an area of 4240 km². The highest elevation of this basin is noted as 440 m (Sushinia hill) and the lowest is 7 m. This basin is composed of six left-bank and six right-bank tributaries.

The entire basin is characterised by highly undulating to level terrain. The Basin may be divided into three parts such as upper, middle and lower based on distinct geomorphological features from west to east, in accordance with the physiographic characteristics.

In the west, the upper catchment essentially belongs to more or less flat terrain with erosional landforms (Chhotanagpur Granite Gneiss Complex). The area is characterised by residual hills, isolated hillocks, tors, sills, dykes, terraces and valley fills.

In the central portion, the Archean metamorphics of the basement landscape grades into the lateritic upland. The middle reaches of the Dwarkeswar watershed are characterised by this lateritic upland mainly covered by forest, valley fills and channel fills and terraces of unique geomorphic distinctions. Similarly, in the lower reaches, the dissected lateritic upland grades into depositional terrain. This depositional terrain is characterised by the presence of river terraces and older meander scars.

Geologically the basin is characterised by six litho-stratigraphic units. These are as follows:

(i) High-grade meta-sediments and meta-basics of Archean era; includes garnetiferous -- sillimanite biotite schist, quartz schist, calcgranulite.

(ii) Porphyroblastic granite gneiss, biotite and quartz, biotite granite gneiss of middle of upper proterozoic (Chhotanagpur Granite Gneiss Complex)

(iii) Ferrugenous gritty sandstone and shale, hard greenish clay with Fe and Ca concretions of Miocene-Pliocene times.

(iv) Laterites, reddish soil with laterite and limonite, laterite debris of middle to late Pleistocene.

(v) Hard greenish clay with Fe and Ca concretions of late Pleistocene to early Holocene.

(vi) Fine sand, silt and clay to sticky plastic clay deposited from early Holocene to recent period.

The largest part of the study area is characterised by alluvial tract mainly due to the result of fluvial action in recent times.

The study area receives an average annual rainfall of 1300 mm. It has been noted that nearly 80% of the total annual rainfall occurs during June to September mainly covering the rainy season. The spatial pattern of rainfall occurrences varies from west to east and south-east. In the western part of the study area the mean annual rainfall is noted as 1300 mm, whereas on the east and south-eastern parts, it is 1500 mm. The temperature condition of the area under reference is very extreme in nature. May is the hottest month with a mean daily maximum temperature ranging from 40°C-43°C and minimum of 20°-22°C. Occasionally, the maximum temperature rises to about 46°C to 47°C by dry land winds. January is the coldest month in this area with mean daily maximum temperature varying from 25°C-28°C and minimum 8°C-10°C. However, due to western disturbances and spells of cold weather temperature goes down to about 5°-6°C.

In general, the study area falls under the moderate morphogenetic region in which subareal denudation mechanisms are very significant. The processes of degradation, particularly weathering, mass movement and erosion are very active in this study area.

A deep level, high degree of weathering has been noted in various parts of the area, which has resulted in the alteration of composition of earth material. Similarly, mass movement and erosion are also very significant in the study area. An exclusive analysis of fluvial process or fluvial erosion in terms of rill and gully erosion reveals that these are also very active in the area. The geomorphic processes and human operations have also resulted in a drastic change in the landmass of the basin in terms of physical, chemical and biological degradation. All three types of degradational processes have been analysed in detail.

Land Degradation Due to Physical Processes

The entire upper catchment or the erosional pediment of Chhotanagpur Granite Gneiss Complex covers an area of nearly 30-35% of the total area. The basin, with occasional exposures of country rocks, hills, dykes and weathered regolith, reflects the effects of physical weathering processes at a considerable scale that are changing the landscape as well as the land stability. The types of weathering that have been noted are rock disintegration due to fluctuation, exfoliation of shells or spheroidal weathering, sheet erosion and erosional off loading, block disintegration, etc. The above degradational processes are mainly noted in Dwarkeswar upper catchment, namely, Beko sub-catchment, Dudhbhariya sub-catchment, Darubhanga sub-catchment, Dangra-Adali sub-catchment and Ghandeshwari sub-catchment which cover parts of Purulia and Bankura district respectively.

Various types of mass movements, considered as one of the important aspects of physical process, may be noted, namely, rock creep, soil creep, soil fall, slumping and bank failure. Lalpur-Majuria hill, Tillabani hill, Dungrigorh hill, Paharpur hill,

Sushunia hill are some distinct geomorphic locations where rock creep and soil creep are very much evident. Soil fall and slumping are mainly observed in the entire lateritic terrain especially in the gully head and gully wall. It includes parts of Bishnupur, Sonamukhi and Bankura, Onda, blocks of Bankura district and areas near Bachaibona and Kashipur village of Kashipur block in Puruliya district. However, some selected area-specific references have been recorded during fieldwork, which are given below (Table 1).

Table 1

Physical Processes and Land degradation: Some Selected References in Dwarkeswar Basin

Nature of Process	Location
Rock disintegration due to temperature fluctuation	Tillabani hill, Dungrigora hill, Paharpur hill and Kapal-kata.
Exfoliation of shells or spheroidal weathering	Tillabani (407 m), Dusgonigora (274 m), Kalabani (240 m), 500 m due south of Amghata village (hbl. – amphebolite) 5km^2 centering Hara, north of Maguria-Lalpur, Chanchipathar, west of Motipur.
Sheet erosion and erosional offloading	500m east of Bahchdiha, 1.2 km due N-E of Ladurka, south of Parasibana, adjacent of Dhanara, north of Saldiha, east of Kapatkata, south of Maguria.
Block disintegration	Prominent example has been noted 3 km S-SE of Maguria at the middle part of 250m hillock (meta-basalt) and 500 m east of Beko bridge (north bank) with two prominent joint planes.
Rock creep, soil creep; soil fall, slumping	Lalpur Majuria hill, Tillabani hill, Sushunia hill. Some areas in lateritic terrain in Puruliya district and in a large scale at Bankura district.

Bank failure, bank erosion, and slumping mechanisms are also significant in the basin. Its rates are noted at high intensity. The evidences of such failure are often accentuated during rainy season or peak runoff period. Lateral spreading and bank erosion are very common along the main stream of river Dwarkeswar near Bishnupur to Arambag. Sequential satellite imageries depict that bank failure and soil erosion are taking place in the right bank covering larger area, rather than the left bank with a tendency to trace the older path (presently marked as channel scar, meander scar).

Land Degradation Due to Water Action

Sheet erosion is prominently noted in the upland areas of the basin, having gentle slope with light textured surface soil underlain by heavy soils. This activity accelerates deforestation in the area. Raindrops fall on the powdered soil, which clog the pore spaces. It ultimately reduces the process of overland flow. The slope, textural characteristics and infiltration rate and rainfall intensity during the monsoon season, definitely increase surface runoff and overland flow which lead to severe top-soil erosion and sheet erosion in tropical lands. Moderate to severe sheet erosion is also significant in the upper catchment of the Dwarkeswar basin. Its evidences are noted at Madhabpur, Hir, Lakkhanpur, north of Nayadih of Hura block and Miraipahari, Amdiha and Barra of Kashipur block.

In the study area the rill and gully erosion are also very common mechanisms. Its evidences are mainly noted in the upland areas of the region. Presence of loamy skeletal or sandy skeletal soils are important contributing factors for excessive soil loss in this study area. It is evident that the slightly undulating terrain of the study area is mainly affected by large-scale development of rills and gullies. The presence of numerous joints, fractures and other weak zones have also accelerated the development of rills and gullies. A large concentration of such type of areas is noted in the Kashipur and Hura block in Purulia district and Chhatna, Gangajalghati, Indpur, Bankura-II, Onda, Bishnupur and small parts of Jaypur blocks in Bankura district respectively. Some selected examples are given below which were recorded during the field visit.

Location	Types of erosion and other associated features
Dhanara	Excessive rill and gully erosion; barren land with scrub; presence of strewn quartz on the surface; gully wall slumping.
SE of Chapora	Severe rill and gully and river-bank erosion.
West of Aguibad	Severe rill and gully erosion; topsoil completely removed, regolith shows full of iron concretions; gully width varies 60-70 cm. and depth 90-120 cm.
SW of Kashipur	Severe rill and gully erosion leading to badland formation; 'sheshol' plantation in the gully wall to check erosion.
South of Dhagora (north bank of Futuri river)	Severe rill and gully erosion; severe dissection resulted 'haystack' landforms.
Rangametia south of Amakunda, East and SE of Katharia-Tilaguri-Lari-Jagannathdihi-Pabra-Haridih-Jibanpur-Lara (a vast tract covering Hura block and Chhatna block)	Severely dissected by rills and gullies; a vast tract of upland where lateritisation processes were active (primary phase); topsoil is nearly removed; Surface is covered with iron concretions and nodules alongwith quartz pebbles and angular fragments of granite. It has been noticed here that mica-schist is intruded by porphyretic granite.
South of Dumrasol	Rill and gully erosion over a long stretch of stony waste.

Soil-Physiographic Relationship and Land Degradation

Physiographic-soil relationship is a widely accepted technique in finding out the land resources (Sehgal, 1990) and potentialities for land degradation. This helps in extrapolation of boundaries using soil-physiographic relationships and it will also help to formulate suggestive measures to protect the land from degradational hazards.

The whole area has been classified into ten physiographic units. Soil association of each physiographic unit has also been established by studying soil profiles and minipits. The soils are classified upto family level following Soil Taxonomy (USDA, 1975). The soil-physiography association and some important soil and land properties are given below.

Physiography	Soil Association	Depth	Surface Texture	Runoff	Erosion Hazard Class
Denudational Hill	L.S., Lithic Ustorthent	d2	Loamy sand	MH	e3
Pediment-Upper (Toe slope of residual hill)	L.S., Typic Ustorthent	d2	Sandy loam	MH	e3
Pediment-Lower (Weathered rocky pavements)	S.S., Typic Ustorthent	d1	Gravelly loamy sand	MH	e3
Buried Pediment Shallow (BPS-Upper slope)	C.L. Typic Ustorthent F.L., Ultic Haplustalf	d2-d3	Sandy loam	ML	e2
Buried Pediment Shallow (BPS-Lower slope)	C.L., Typic Rodustalf F.L., Ultic Haplustalf L.S., Ultic Haplustalf	d3-d4	Sandy loam	ML	e3
Buried Pediment Medium	C.L., Typic Ustochrept F.L., Aquic Haplustalf F, Vertic Ustochrept	d5	Loamy sand Clay loam Sandy loam	ML H ML	e2
Dissected Lateritic Upland	L.S., Typic Plinthustalf L.S., Typic Ustorthent (slope area)	d3-d4	Sandy loam Gravelly sandy loam	MH ML	e3
Older Alluvial Plain	F., Typic Epiaqualf F.L., Typic Epiaqualf F.L., Typic Epiaquept	d5	Clay loam Sandy loam Sandy clay loam	ML	e1
Younger Alluvial Plain	F., Typic Epiaquept F.L., Typic Epiaquept	d5	Clay loam Clay loam	L-ML	e1
Infilled Valley	F.L., Typic Epiaquept F.L., Aquic Ustochrept	d4-d5	Sandy clay loam	ML	e1

L.S. – Loamy sand; S.S. – Sandy skeletal; C.L. – Clay loam; F.L. – Fine loamy
d1 – Very shallow (0-10 cm), d2 – Shallow (10-25 cm)
d3 – Moderately deep (25-50 cm), d4 – Deep (50-100 cm), d5 – Very deep (100 cm)
L – low, ML – Moderately low, MH – Moderately high, H – High
e1 – Slight, e2 – Moderate, e3- Severe

Evidences of chemical weathering are observed in the study area in different modes and at different scales. In case of rock weathering, as already mentioned, some examples of spheroidal weathering may be noted in and around Tillabani, Dungrigora and Kalabani hills and also at various places around Hura. Although spheroidal weathering is commonly supposed to be due to expansion, but the layers in a spheroidally weathered block are caused by chemical migration of elements within the rock, with periodic deposition of the Liesegang ring type (Ollier, 1969).

Major portion of the Dwarkeswar Basin being most primitive landmass of Indian subcontinent, i.e., Chhotanagpur Granite Gneiss Complex of Pre-Cambrian era, has evidences of chemical weathering in the development of soils and alteration of profiles which are also very well significant in the region.

Colours are important indicators for chemical weathering. Red, brown and yellow are commonly attributable to chemical weathering. Red ground is generally attributed to non-hydrated iron oxides and yellow ground to hydrated iron oxides or to mixtures of iron and aluminum oxides (USDA, Soil Survey Manual, 1951).

Intensive process of chemical weathering is noted in the soils associated with denudational hill (brown), toe slope of residual hill (brown to reddish-brown), isolated pediments or weathered rocky pavements (dark red to brown), upper slope of buried pediment shallow (brown to strong brown, and yellowish-brown to brown associated with Ultic Haplustalf), lower slope of buried pediment shallow (reddish-brown to red and yellowish-brown), and dissected lateritic upland (strong brown to yellowish-red).

Product of deep level weathering is very important to understand the intensity and process in the upper catchment as well as in the lateritic upland of the study area. In this connection, occurrences of china clay are considered as special reference, which are very common in this area. Here, china clay is associated with highly weathered profile of phyllites of the iron-ore series and granite rocks. Examples of such occurrences are also noted around Dhatra (23°37′N; 86°43′E), Kalajhar (23°27′N; 86°35′E), Rangamati

(23°23'N; 86°50'E), Taldih (23°24'N; 84°56'E), and Tundi (23°59'N; 86°023'E). The decomposition of feldspathic constituents in the granitic rock has given rise to kaolin. It is found to a depth of 10-15 m from the surface. The parent rock, on washing, yields 8-10% of pure china clay. Chemical analysis of a typical china clay sample has yielded the following results:

SiO_2 -- 48.35%	TiO_2 -- 0.25%	MgO -- 0.98%
Al_2O_3 -- 30.75%	CaO -- 10.21%	Na_2O -- 1.65%
Fe_2O_3 -- 2.24%	K_2O -- 0.28%	

Anthropogenic Processes

Multifarious man-induced activities have resulted in various types of hazards which have degraded the land of the region. These activities are mainly due to unplanned terracing and agricultural activity on the slope, intensive grazing, deforestation, extraction of surface materials, etc.

Conservation and Management Proposals

Considering various types of land degradation and its severity, conservation approaches and management proposals for sustainable land management are imperative. Various procedures of land treatment are given below considering the mechanical, vegetative and forestry approaches:

- Denudational hills under different physical and chemical weathering processes may be brought under afforestation with staggered contour trenching. Conservation and upgradation is required for existing forest areas.
- In the stony waste, social forestry may be approached through plantation of Akashmani (*Acacia auriculaeformis, Acacia modesta, Acacia seiberiana*) and Banaraj (*Bauhinia racemosa*).
- Severely sheet-eroded land, with top-soil completely removed, existing as land without scrub, may be recommended in the isolated pediments for plantation of Akashmani (*Acacia* sp.) and Eucalyptus (*E. citniodora, E. globalus, E.odorate, E. occidental*). In the buried pediment shallow (BPS) areas, apart

from above mentioned vegetative treatment, silvi-pasture [Subabul tree (*Leucaena glauea*) + Stylosanthes (*Stylosanthes gracilis*), Dinanath (*Pennipetum pedicillatum*), Maize (*Zea mays*)] may be recommended.

- For gully and rill erosion Khas Khas (*Vetivenia zizamioides*), Kher or Saqri (*Heteropogon contortus*), Kharang (*Aristida* sp.) may be planted with Sisul (*Agave sisalava*). Sabai grass (*Eulakopsis binata*) may be grown in/across the gullies or bordering wastelands to check land degradation.

- In the areas of BPS lower slope, marginally single crop area with low productivity hazard potentialities in moderate to severe. Proper land treatment and suitable agricultural practices should be given prime importance. Soil conservation measures like terracing, contour ploughing and improved crop management practices -- crop rotation, strip cropping -- may be undertaken. Amongst others, agro-forestry and agro-horticulture are considered to be most suitable practices for better sustainable land management. Agro-forestry like Teak, Sissoo, Palas, Mahua, Kendu, Neem along with kharif crops (mixed farming) may be undertaken. In kharif season, mixed farming like paddy+arhar, groundnut+arhar, soyabean + arhar may be practiced. Agro-horticulture (in areas of good road network) like Guava (*Psidium guagava*), Amla (*Emblica officinalis*), Karanda (*Carrisa caranda*), Kul (*Zyzyphus jujuba*) in association with kharif cultivation should be undertaken in the developmental programmes.

- Identification and mapping of fluvio-geomorphic features like derelicted channels, older channel scars, meander scars and scrolls, ox-bow lakes, younger depositional terraces, etc., is very essential to understand their significance in hazard assessment. Bank failure/bank erosion, sand splaying, etc., are also significant in this regard. Accordingly, suitable structural engineering measures like embankments, sluices, spurs etc., may be constructed. Proper vegetative treatment is recommended along the bank line to check severe soil loss during the flood and post-flood periods.

However, the basin being very typical in shape, i.e., semicircular in the upper catchment and elongated in remaining parts, characterises flash flooding in the partly upper and middle part of the basin. An immediate programme for sustainable watershed management is required for the socio-economic development of the region.

References

Saha, S.K., Singh, B.M. (1991). Soil Erosion Assessment and Mapping of the Aglar River Watershed (U.P.) Using Remote Sensing Technique, *Journal of the Indian Society of Remote Sensing*, Vol.19, No.2, 67-76 pp.

Saxena, R.K., Verma K.S. and Barthwal, A.K. (1991). Assessment of Land Degradation Hazards, Etah District, U.P. using Landsat Data, *Journal of the Indian Society of Remote Sensing*, Vol.19, No.2, 83-93 pp.

State Remote Sensing Centre, DST and NES, Government of West Bengal, Calcutta, December 1997, Project Report on Soil Mapping of Dwarkeswar Watershed, Purulia District, West Bengal, IMSD Phase-II.

State Remote Sensing Centre, DST and NES, Government of West Bengal, Calcutta, July 1999, Report on IMSD Phase-II Project, Dwarkeswar Watershed, Bankura District, West Bengal.

Ollier, C.D. (1969). *Weathering*, Oliver and Boyd, pp. 1-45.

20

Land and Natural Resources Tenure Reform in Developing Countries: Lessons for Zambia

Gear M. Kajoba

Land reform in many developing countries has attempted to ensure that small-holder farmers get access to land, water and other natural resources, and that people, especially women, are integrated and participate in rural development. It is hoped that in this way sustained food production can be achieved to meet the requirements of increasing rural and urban populations, so that agriculture can contribute to the growth of the modern economy.

Increasing population tends to exert pressure on natural resources, especially forests, as there is need to clear land for cultivation and harvest forest resources such as timber for the modern economy and charcoal and firewood for domestic use. Such pressures have necessitated the need to devise sustainable management strategies which ensure the participation of local communities and all the stake holders to arrest deforestation and promote agroforestry systems and tree planting.

In most rural societies, women play a leading role in food production and the management of the household economy. However, there is still a tendency which results from the operation of traditional norms, to marginalise women and prevent them from adequately accessing factors of production such as land, and constraining their contributions to the management of natural resources such as forests.

This paper which is based on the review of literature therefore attempts to review current trends in land reform and the management of forest resources, and recognises the role of women in tree and forest management in developing countries, in order to establish what lessons Zambia can learn from experiences elsewhere. The paper is structured as follows: section 2 gives an overview of the global situation with respect to land reforms, forest management and the role of women in tree and forest management; section 3 examines the situation in sub-Saharan Africa, while lessons for Zambia are drawn in section 4; and section 5 is the conclusion.

The Global Situation: An Overview

Land Reform

According to Herrera et al (1997), most of the land reforms which took place in Latin America, Asia and the Near East after the Second World War were modelled on the experiences of Russia. Following the revolution of 1917 a variety of populist and socialist oriented regimes moved the ideology of land reform from a liberal economic process to a state engineered way to redistribute land and achieve equity in rural areas. During these land reforms it was considered that the state rather than the market determined redistribution of land. Land reform was seen as one of the main tools for achieving rural transformation and agricultural development.

In Latin America, land reform especially since the 1950s has dealt with the complex issue of how to redistribute land from the powerful landlords to the peasants in the hacienda system, often referred to as Latifundio/Minifundio complex (Kay, 1998). Starting with the Mexican Revolution of 1910, efforts at land reform, which emphasises redistribution of land, have taken place with varying degrees of success.

In some countries the integration of peasants into the national economy and institution building, have been seen as ways of encouraging participation and stability in the countryside (Kay, 1998), although it has been argued elsewhere that the reform process was ineffectual and has been neutralised by lack of follow-up measures such as the provision of inputs (Thiesenhusen, 1995).

The significant development since the 1980s in Latin America is what Kay (1998) refers to as the neo-liberal challenge. By this he means that the neo-liberal winds of change have led to the privatisation of credit, marketing and other technical assistance which was provided to the rural sector as part of state subsidised services. This neo-liberal challenge has been of advantage to commercial farmers who were able to increase exports with the renewal of price controls, but a disadvantage to peasant farmers. However, some peasant groups with better endowment, entrepreneurial skills, locational advantages in terms of closeness to markets and with access to external assistance from non governmental organisations (NGOs), have adapted successfully to this challenge (Kay, 1998).

In Colombia, for instance, the new land reform law of 1994 aimed at the promotion of market-assisted land reform. It involves stimulating the participation of previous agrarian reform target beneficiaries in land negotiations.

Similarly, land markets have opened and land transactions are being registered in the more economically dynamic coastal zone of Ecuador, where export crops such as cocoa, bananas and sugarcane have been grown since the colonial times (Herrera, et al, 1997). In Peru, parcellation of haciendas and the co-operative farming in the reformed sector has led to the growth of new peasant landowners. Even in Cuba, the demise of the Soviet Union and the transition of the former socialist countries from a planned to a market system, has led to some reforms. Greater economic incentives have been provided for peasant farmers and producer co-operatives. In 1994, private agricultural markets were introduced where the state no longer controlled prices and "where producers can sell any surplus production that remains when they have met their quota for the state market" (Kay, 1998, p.27).

Thus, the trend in land reform in Latin America is that the World Bank and NGOs have financed programmes of land registration and titling and a new land policy that emphasises free markets and security of property rights in order to stimulate long-term investments. Kay (1998) concludes that despite the continuing arguments by scholars and activists in favour of land reform, "the era of radical agrarian reforms, however, is over". Instead, "there has been a shift from state-led and interventionist oriented reform programmes to market-oriented land policies" (Kay, 1998, p.28).

In Asia, both China and Vietnam are undertaking land reforms which intend to give farmers more flexibility to undertake individual initiatives and enjoy relative security of tenure. Since the 1970s, China has undertaken land reforms which have departed from the collectivisation of the 1950s. These reforms have seen the introduction of the family based contract system known as the responsibility system (Fu and Davis, 1998). In this system, farmers are given freedom of landuse rights and decision-making, linking rewards closely with their performance. While the state is still the owner of the land, farmers are being given leases. By 1995, leaseholds were extended from 15 years to 30 years (Fu and Davis, 1998, p.128).

These changes have given rise to a debate among economists as on how further reforms should be carried out. While the first group favours nationalisation of land, and giving permanent landuse rights to peasants, the second group is advocating privatisation of land with individual ownership. The third group is advocating for a redefinition and improvement of landuse rights of the peasants. These argue that ownership of land was just one component of property rights. Other components include the rights to consume, to sustain income from and to alienate assets. (Fu and Davis, 1998).

It is important to emphasise that these land reforms which have given the Chinese peasant relatively more secure landuse rights at the family level, have contributed to accelerated agricultural production after three decades of stagnation. It is reported that grain increased at annual rates of 4.8%; cotton at 7.7% and oil-bearing crops at 13.8% (Fu and Davis, 1998, p.124).

Since the end of the cold war, Vietnam is undergoing a process of transformation from a centrally planned to a market-oriented economy. Decollectivisation of agricultural production organisations has taken place, and land has been allocated to private individuals and households for stable long-term use (Haque and Montesi, 1996). The land law of 1993 gives security for periods of 20 years for annual crops and 50 years for other perennial crops and forestry. Also, the law permits the leasing of land to foreign investors. It is stated by Haque and Montesi (1996) that these tenurial reforms have resulted in more intensive use of land and impressive growth of agricultural output, which have transformed Vietnam from a food-deficit country into a food-surplus country.

It may be necessary at this point to make a brief reference to the situation in Eastern Europe where the transition from centrally planned economies to market orientation started. Micheva (1994) points out that in Bulgaria, a process of restitution in which land ownership is restored to its original owners is underway. It is expected that 2 million agricultural holdings will be restored to their pre-Sovietisation owners. When this process is completed, one in every four or five or 20 to 25% of the Bulgarian population will possess an estate or will become farmland owners. Thus, many new small-scale family farmers will be created.

Forest Resource Management

Most of the literature on the management of forests in developing countries, has tended to focus on the efforts being made by India and Nepal in Asia. In both these countries, the closely related concepts of community forestry and joint forest management (JFM) are being implemented. These concepts entail the forging of a partnership between the state and the local communities in order to manage forest resources in a sustainable way. Campell (1995) is of the view that JFM envisions sharing the responsibilities for the protection and management of forest resources, as well as the benefits which are realised from these forests.

According to Palit (1995), JFM began in India in 1988, and its common objective "is to secure the willing co-operation of the people through their active participation for the conservation and development of forests on a sustainable basis" (Palit, 1995, p.25). It requires that both industrial demands (for constructional timber, plywood, paper, etc.) and local demands for housing material, fuelwood, energy, fodder and cash, are taken into account.

Where JFM has been established and taken root, such as in West Bengal, numerous community groups in the form of forest protection committees or forest user groups sprung up. These groups participate in planning, implementation, monitoring and evaluation (Choudhury, 1995). The user groups thus, make micro-plans which make an assessment of resources, community needs and focus on the central issue of why, how, by whom and for whom the forest itself should be managed (Choudhury, 1995, p.189).

The user groups frame rules and regulations on the harvesting of forest resources, including the collection of dry fallen twigs and branches. While in some instances the collection of these may be totally restricted, in others, even non-members of the user groups or forest protection committees may be allowed to collect dry twigs. Whereas grazing is closed in JFM areas in West Bengal, afforestation is also undertaken in Gujarat, and the groups mount patrols on a daily basis to enforce the regulations. In terms of sharing, the forest protection committees are entitled to 25% of net proceed in W. Bengal, whereas in Gujarat it is 50% (Dobriyal and Ballabh, 1995, pp.221-222).

It is indicated that some complex situations have arisen in the process of managing JFM forest areas. Although the unit of forest protection in JFM is a part of forest comprising several villages, there are instances where people from outside villages are accused of misusing the forest. In addition, conflict has also arisen within a forest protection committee, as some villagers do not take as much interest in forest management as others (Dobriyal and Ballabh, 1995).

Thus, through JFM many hectares of once degraded land are now regenerating in some parts of India, reversing the negative trend of decades (Campell, 1995, p.7).

In a personal communication with Dr. Yam Malla (13.4.2000), it was learnt that community forestry in Nepal began in the 1970s, and was a result of international concern for environmental degradation in the Himalayas. Degraded forests were handed over by the government to local communities for reforestation. Community based forest user groups prepare management plans which have to be approved by forest officers, and the groups reap the benefits which are used for community concerns.

Gilmour and Fisher (1991) state that this policy of handing over of all accessible hill forests to be managed by local communities, was seen as a "bottom-up" approach to forest management, which resulted from a shift in the existing paradigm. Local communities were now perceived as being at the centre of the "forest problem" and their indigenous knowledge could be tapped in managing the resources. The new approach was radically different from the previous one in which only foresters were considered qualified to manage forests.

Thus, a well-developed agroforestry system has emerged in Nepal in the past twenty years or so. Already existing local institutions (such as forest watchers) have been mobilised in conjunction with state sponsored ones, to manage forests. Farmers have also planted and managed trees on private farm land as a response to a growing scarcity of forest products from community forests (Gilmour and Fisher, 1991, p.65). New perceptions on community forestry have arisen. It is increasingly being argued that while the control of forests by local communities in India, Nepal and to a certain extent the Philippines has been addressed, there is need to go beyond resource conservation and subsistence use, and consider how individuals, families and communities can derive better incomes from the management and extraction of forest and tree products (Victor, 1997).

Byron (1997) argues that in the past there was some naivete in the expectations that communities would participate in managing forest areas with little prospect of tangible benefits. He emphasises that individuals, communities or sub-groups within communities "do not engage in forestry primarily because it makes them feel good about helping others; they make the effort because they see it worthwhile to themselves and their families, and this often has a financial element to it" (Byron, 1997, p.4).

Thus, emphasis is now being placed on income generation from the sale of forest products, fees, fines and donations from donor agencies. The income is playing an important role in creating local employment and developing local markets (Hunt et al, 1997, p.64). Marharjan (1997) contends that in the Koshi hills of Nepal, subsistence forest user groups, particularly the poor and women, are generally more motivated to conserve forest areas when there are quick returns from such activities. At this stage, it may be helpful once again, to make reference to the new experiences in forest management which have emerged in the countries of Central and Eastern Europe (CEE). El-Lakany (1997) points out that dramatic shifts in forest ownership patterns have taken place. Through the processes of compensation, restitution and privatisation, thousands of state forest land have been given to private individuals. The number of private forest owners with small and sometimes dispersed holdings has increased. This means that new forest extension strategies have to be devised in order to take into account the needs of small private owners. While previously, government personnel managed the forest estate, privatisation will entail management of individual forest holdings by the owners themselves.

In countries such as Croatia, Czech Republic, Hungary, Lithuania, Poland, Romania and Slovenia, the overwhelming majority (over 90%) of private forest owners have a property size under 5 hectares. In Croatia, Poland and Romania, 100% of forest owners fall under 5ha size (Marghescu and Anderson, 1997, p.15). Although there are temporary restrictions for selling restituted land, especially to foreigners, an informal market for privatised forest land with illegal or unofficial trading does exist. In order to raise capital and modernise technology, forest industry enterprises which have also been privatised, have established joint ventures with foreign investors. Agroforestry programmes and the conversion of marginal degraded or abandoned agricultural land into forests, have to take into account tenure changes which have taken place (El-Lakany, 1997; Marghescu and Anderson, 1997, pp.15-16).

Women and Forest Management: Trees and Tree Tenure

In many rural households in Africa, Asia and Latin America, women have gathered forest products for fuelwood, for fibre, mat-making and basket-making; for food as vegetables, nuts, fruits, vines and medical uses (Dankelman and Davidson, 1988). Women also collect animal fodder in addition to other duties in the household (Smyth, 1999).

Despite increasing deforestation caused by the conversion of forests to agricultural land and due to commercial timber logging, women are still expected to gather wood and other biomass for fuel wherever they can. They have to take twigs, smaller branches and dead wood.

In addition, many rural women obtain income from small-scale, forest-based enterprises. These include traditional crafts, bee-keeping, making mats, selling herbs, weaving baskets and other crafts and selling fuelwood. In India for instance, small-scale forest-based enterprises provide more than 60% of forest-based exports and are responsible for 1.6 million of the 2.3 million jobs or about 70% of jobs, in the forestry sector as a whole. About 600,000 women and children collect or harvest more than 350,000 tonnes of leaves from the Tendu tree, which are used for wrapping tobacco to make a small cheap cigarette called a *beedi*. A total of 2.5 million people were involved in *beedi*-making in India in the mid 1970s (FAO and SIDA, 1991).

Women are not only suffering from the effects of deforestation which means that they have to cover longer distances to fetch fuelwood (with older women carrying heavy bundles of wood) (Smyth, 1999), but they are also responding by leading attempts to reverse the destruction by planting trees. All over the world, "women have reacted either spontaneously or through local organisations, to protect forest resources from destruction and to ensure that future supplies will be adequate" (FAO and SIDA, 1991, p.25).

In India, for instance, the Chipko movement protests against forest destruction and organises eco-development training camps in which tree planting is promoted. The survival rate of plants in Chipko plantations, which are cared for by local women, are between 80 to 90%. Usually, the species which are preferred are those which provide fuel and fodder (Dankelman and Davidson, 1988; FAO and SIDA, 1991). While women in Thailand are involved in settled agroforestry systems, those in South Korea are playing an important role in the programme to re-green the countryside (FAO and SIDA, 1991). In the Philippines, women are successfully participating in schemes to grow trees as cash crops. In Pananao, which is located in the central mountains of the Dominican Republic, women play different roles in comparison to those of men in the management of forest resources such as palm trees. Exploitation of the palm tree varies according to the parts of the tree being used. While men use the wood for construction, women use the fibre in handicrafts, and both men and women use the tree's products for animal fodder. The division of control becomes more complicated with change in the location of the trees. Near the homestead, only women are involved in the use of the resource, but on pasture land men exercise all the functions except for the collection of fuelwood. On cropland the division of labour is mixed. On forest 'remnants' women are given responsibility for the trees and provide the labour, but men have control of the palms. However, women control all processing activities even though they do not manage the source areas of the new materials they need, many of which come from the men's fields, pastures and woodlands (FAO and SIDA, 1991, pp.14-15).

Despite women's dependence on forest resources, forestry policies have tended to ignore women and their activities. Even social forestry programmes do not include women in sufficient numbers. Women's participation is constrained by many factors such as: problems over land tenure and an absence of legal entitlement to land; lack of time left after domestic duties; cultural taboos and lack of training or familiarity with forestry and other institutional support (Dankelman and Davidson, 1988).

Women need security of tenure in order for them to benefit from forestry projects and forest management. Chambers and Leach (1987) argue that tree rights for many poor people in Asia, especially women, tend to be ambiguous within the context of social forestry projects. The participants in these projects do not own the trees they have planted. Even where trees are on their own land, "they are often prohibited by law, and impeded by bureaucracy, from cutting them down when they want to" (Chambers and Leach, 1987, p.16).

Similarly, Bruce and Fortmann (1988) contend that land and tree tenure are crucial in determining whether those who participate in agroforestry programmes can benefit from them. It is therefore crucial that women should have rights in the trees and control the land on which such trees are planted in order that agroforestry programmes be sustained. Women without land cannot plant trees and those who are denied the right to use certain trees have no incentive to plant them (FAO and SIDA, 1991)

The Situation in Sub-Saharan Africa

Land Reform and Agrarian Transformation

With the exception of South Africa and Zimbabwe, where land reform involving land redistribution, restitution and resettlement is on the agenda (Binswanger and Deininger, 1993; von Blanckenburg, 1994; Ntsebeza, 1999), the debate on this issue in most of sub-Saharan Africa is concerned with the need to harmonise traditional communal tenure with modern individualised tenure, in order to create an enabling environment for agricultural transformation and sustainable development.

Chinene et al (1998) contend that there is a general agreement that traditional land tenure rules in Sub-Saharan Africa do not provide the necessary enabling environment for agricultural development because land rights are not registered, and are thus not sufficiently secure for long-term investment. This state of affairs exists at a time when land is becoming scarce due to rapid population growth and increased commercialisation of agricultural activities.

Plateau (1992) points out that property rights economists are of the view that there is need to create freely tradable private property rights which could be the basis for modernising traditional agriculture in Africa. This new approach must allow a free land market to develop in order to induce a switch from subsistence cultivation to commercial agriculture, and also allow for the emergence of dynamic agricultural entrepreneurs who are adaptable to technological changes and evolving market conditions.

It may be argued at this point that the tendency towards individualisation of tenure has its roots in the pre-colonial and colonial periods in some parts of Africa, such as among oil-palm producers in Senegal, cocoa farmers in Ghana and the Chaggaa coffee farmers in Tanzania (Plateau, 1992; Fernandes et al., 1989).

This is an indication that the agrarian systems in sub-Saharan Africa were evolving from the communal types to the more advanced ones and entrepreneurs were able to respond positively to economic opportunities available to them at that time. Although communal land ownership and subsistence agricultural production are still prevalent, there is no reason why transformation should not occur when the conditions are ripe or when it has become imperative to do so (Hyden, 1986).

With the implementation of Structural Adjustment Programmes, the reform of traditional tenure is now on agenda in many sub-Saharan countries. In Burkina Faso, the 1991 Agrarian and Land Tenure Reform Law declared that land belonging to the national estate could now be assigned as private property. This gave investors new possibilities (Faure, 1995). In some remote rural areas of Burkina Faso an informal land market exists in which land can be exchanged for money. In the lowlands where market gardening, tree cropping and rice growing occur, local heads of family who control the best land are able to trade it. They can also loan out their plots, thus creating a clientele of debtors (Speirs and Marcussen, 1998).

According to the Land Act of 1998, a number of reforms have to be undertaken in Uganda. For instance, all Mailo land (land granted to Bunganda in the Uganda Agreement of 1900), was to be converted to freehold land; customary tenants on public land were to apply for freehold; and all leases on public land were to be converted to freehold. The 1995 constitution recognised customary land and made provision that all citizens owning land under customary tenure may acquire a certificate of customary ownership; and this certificate may be leased, mortgaged and pledged, where the customs of the community allow (Mwebaza, 1999). Thus, the customary certificate holder now has "the right to lease the land or any part of it, subdivide the land....transfer the land....or dispose of the land by will" (Mwebaza, 1999, p.5).

However, Nsamba-Gayiiya (1999) cautions that in order to implement the land reforms in Uganda, there is need for flexibility over time. She argues that the process should avoid going too fast; it should be carefully thought out and be sustainable, and all institutional stakeholders should be fully consulted and involved.

In West Africa, Delville (1999) suggests that efforts should be made to remove legal pluralism, as this causes uncertainty over rights in land. He shows that there is need to give legal status to all the local rights as is being done under the Rural Land Plan in the Ivory Coast and the Rural Code of Niger which takes into account local farming, pastoral and forestry practises. But, he warns against imposition of private land ownership especially in areas where land value is low or tenure is so secure that rural communities see no need to register land rights and transfers. He adds that private ownership could come about through "gradual evolution of existing rights facilitated by the law (and) the paradigm is thus one of "adaptation", rather than "substitution" " (Delville, 1999, p.16).

In this regard, Bruce (1993) points out that the tenure reform debate tends to have ideological connotations in terms of those who wish to see either capitalist or socialist transformations of Africa. But given the transformations that have taken place in the former Soviet Union and Eastern Europe, it could be argued that there is need to consolidate the market stage through the promotion of commercial agricultural production by small-holders. Funnell (1991) argues that small-holders who adopt husbandry methods to produce cash crops such as cotton and tobacco on communal land as part of a spontaneous process of commercialisation, "represent the vanguard of petty capitalism and a critical feature of the agrarian structure" (Funnell, 1991, p.176).

Forest Management

Literature on forest policy and management in sub-Saharan Africa outside of government sources is scanty. However, it would seem that efforts are being made in some countries to transform the management of the forest estate from being the sole domain of the forest department, to promoting community participation through joint forest management (JFM). The change also entails the empowerment of individuals and communities through tree planting in agroforestry projects.

According to Scherr (1997), recent support for agroforestry and social forestry is justified on the basis of their potential contribution to rural welfare, which includes subsistence needs such as food, fuel, building materials; as farm inputs for live fencing, animal fodder and green manure; supplementation of cash income through the sale of the products; and social uses such as shade, privacy; and shrubs may be used to rehabilitate eroded or degraded soil and water resources, and timber trees may be used as household savings for the rural poor (Scherr, 1997; Chambers and Leach, 1989).

In Tanzania, Ssembajjiwe (1997) shows that village communities have started participating in joint management of forests. He indicates that in the recent past the Duru Haitemba Miombo woodland forest in Babati district has suffered rapid degradation and encroachment. This prompted the local people to take over the management of this open resource.

By 1995, eight villages established village owned and managed forest reserves over the various areas of Duru-Haitemba forest. Each village prepared management plans to guide conservation, use and management of their forest reserves. By-laws were made which became the basis for governing the use of the forest, and were approved by the local district council. Villagers are issued with permits to allow them access to the forest resources. Enforcement of the by-laws is carried out through intensive patrols undertaken by the village youth. Records are kept for offences and outstanding fines. Elders and leaders in the community command a great deal of respect, and all by-laws are strictly adhered to and complied with. Conflict between different stakeholders (i.e., the forest department, local administration and the community) is minimised by working closely together. Changes in forest policy and management can also he explained in terms of the concept of 'induced innovation'. In this case small-scale farmers are not only interested in 'welfare maximisation', but they have to respond to historical changes such as population density which tends to exert pressure on available forest resources, and to market development which influence forces of supply and demand and prices (Scherr, 1997)

Among the Luo of South Nyanza district in Kenya, deforestation and permanent settlement sharply reduced off-farm land and tree resources by the mid 1900s. Increased population meant higher demand for wood products, and in response, farmers began to domesticate local tree species and adopt the planting of exotic timber species in the underused parts of the farm, around homesteads, or along field pathways or borders (Scherr, 1997).

After independence, agroforestry practices have intensified in Kenya as a result of the ongoing transition to permanent cropping which is associated with individualisation of land tenure, the disappearance of communal tree resources and the rise of local cash markets for fuelwood, poles, seedlings and fruit (Scherr, 1997). By 1990, CARE International's Agroforestry Extension Project worked with 3,000 farmers in 280 farmers' (mainly women) groups and 3,000 primary schools. The project shows that farmers are willing to intensify their level of the husbandry in order to address household needs for food, fuelwood and construction timber and cash income. Inter-cropped trees include Leucaena or Sesbania Sesban (for green manure) while Eucalyptus and other species are grown for the urban construction pole market. In addition, fruit trees such as citrus and papaya are grown as food for children especially during the pre-harvest period (Scherr, 1997).

Table 1
Types of Agroforestry Projects in Selected African Countries by 1990

Country	Year Project Started
Type A: Tree growing to increase productivity	
Rwanda	1970s
Tanzania	1984
Cameroon	1984
Type B: Tree growing for fuelwood	
Rwanda	1984
Kenya	1983
Zimbabwe	1983
Ethiopia	1986
Type C: Village forestry projects (woodland)	
Mali and Buriana Faso	1977
Tanzania	1980s
Senegal (tree planting around wells)	1975
Type D: Tree planting in dryland areas	
Niger (windbreak project)	1975
Mali	1983
Burkina Faso	1979
Zambia	1986
Type E: Promotion of natural regeneration	
Tanzania (soil conservation)	1973
Kenya	1981
Niger	1980

Source: Adapted from Kerkhof, 1990-list of projects covered in survey.

Agroforestry projects are also underway in different parts of sub-Saharan Africa. Table 1 summarises the different types of agroforestry projects which were surveyed by Kerkhof (1990). He found that significant increases in tree growing occurred where farmers were provided with seedlings or seeds of the species they wanted. While new species and cultivation techniques have been adopted, small-scale farmers are reluctant to adopt radical changes in farming techniques especially if these require high inputs of labour. Farmers prefer small, gradual changes in their farming methods. In dryland areas, it was found that tree growing was much more difficult as growth rates are slower, survival rates are poorer and seedling protection is more of a problem. Here, encouraging natural regeneration is more acceptable to farmers than tree planting because it is cheaper and less risky. Furthermore, it was found that in pastoral areas, tree-planting efforts have had poor results as costs are high while growth rates are usually very low. Kerkhof (1990) argues that protecting areas from grazing animals is a much more effective way of restoring the natural vegetation and tree cover, although control over grazing rights is a very sensitive subject in any pastoral community.

With respect to incentives in agroforestry projects, Kerkhof (1990) found that farmers in Africa are interested in growing multi-purpose trees as sources for construction wood, poles, fruit and other products, and especially as a source of cash income. Other uses include shade and the marking of farm boundaries. He emphasises that projects which focused only on the provision of fuelwood have faced problems.

Similarly, Timberlake (1988) points out that most of the attempts to persuade farmers in the drought-prone Sahel region to plant trees for fuel have ended in failure, because people prefer to plant trees for multiple purposes. Trimmings from such trees are then used to meet fuel needs. He further argues that lack of clear land ownership in many African countries destroys any incentive for farmers to invest in long-term enterprise of tree growing or in the conservation of soil and water. Where communal tree growing programmes have been undertaken in the Sahel, farmers were not sure whether they would benefit as individuals. Some expected the trees to go to the government forestry service, or to the village chief or to the donor agency that organised the project.

It is therefore imperative that tree and land tenure should be clearly spelt out in agroforestry programmes. Where individuals know that planted trees belong to them, they will have a greater incentive to plant more (Thomson, 1988; Fortman and Bruce,

1988). Furthermore, it is argued that in parts of Kenya, such as in the Mbere region where individualisation of tenure has taken place, people's perceptions on land and trees have changed radically. Firewood is no longer perceived as a 'free good' as individuals are now excluding others' rights over their land. Land and trees are now privately owned (Brokensha and Riley, 1988). In Lesotho, private ownership of trees is complete where trees are grown on parcels of land on which individuals have secure rights (Turner, 1988).

Women, Agroforestry and Trees

According to Fortmann and Bruce (1988a) little is written about women and trees, despite the fact that women play important roles in forest management and the utilisation of trees. In many countries of sub-Saharan Africa women participate in agroforestry programmes although they do not enjoy secure rights to the planted trees because their rights in land on which the trees are planted are rather insecure. In this regard Rocheleau (1988) has observed that even countries where both customary and legal reforms have provided for equality, the marginalisation of women still exists with respect to use and ownership of agricultural land, and this situation works against the adoption of agroforestry technologies by women.

Obi (1988) contends that women in Nigeria enjoy two types of interests in economic trees, i.e., 'direct' and 'derivative' rights. Direct rights are obtained irrespective of their status as a wife; while derivative rights and interests are those which a woman may enjoy by virtue of the legal position as a wife, widow or mother. Thus, among the Ibo of Nigeria, a woman may plant fruit trees such as Paw Paw and Plaintain in her matrimonial home, and is allowed to purchase trees either with or without the land on which they grow. She can also lease a tree or take a pledge or accept a gift of them.

However, women's rights to land and trees are often tied to marital status within the context of customary and statutory law, and these rights tend to be eroded during the translation of custom to 'modern' systems.

Warner (1997) points out that in many societies of Eastern Africa, women are recognised as users not owners of (forest) resources. Since men control resources, they also exercise the right to make decisions about planting trees. According to Bradley (1991), research conducted among the Luyia people of Kakamega district in Kenya, showed that tree-planting activities were dominated by men. The ownership of trees by men has been effectively sustained through well manipulated cultural taboos, such as the belief that if a woman plants a tree she will become barren and the husband will die (Bradley, 1991, pp.206-207). This state of affairs is most unfortunate in view of the fact that tree planting in woodlots has increased in Kenya especially on individual private land where tenure is more secure, and the Green Belt Movement by Kenyan women is very active (Dankelman and Davidson, 1988).

For many rural women, especially in the Sahelian states, agroforestry concepts of combining trees, shrubs, palms, bamboo, with agricultural crops and sometimes animals, are traditional. According to Dankelman and Davidson (1988), the typical landscape of Sahelian countries is "farmed parkland" – individually managed fields, pastures and fallows and dotted trees. Villagers combine crops with trees such as Shea-nut and baobab.

Lessons for Zambia

Land Reform and Agrarian Change
We have noted in the literature review above that since the collapse of the former Soviet Union and transformations in Central and Eastern Europe, land reform in Latin America and Asia has been greatly influenced by the neo-liberal challenge. This has entailed acquisition of land through free markets rather than through state redistribution. It has also led to privatisation of credit, marketing and other technical assistance which was provided to the rural sector as part of state subsidised services.

We have further noted that with the exception of South Africa and Zimbabwe where land reform involving land redistribution, restitution, reform, and resettlement is on the political agenda, the debate on this issue in most of sub-Saharan Africa is concerned with the need to harmonise traditional communal tenure with modern individualised tenure, in order to create an enabling environment for agricultural transformation and sustainable development.

With respect to the above trends, Zambia seems to be on course as she has also embraced the winds of neo-liberal challenge since 1991 through liberalisation of agricultural marketing, removal of subsidies, and the enactment of the Lands Act of 1995 which among other things makes provision for the conversion of customary tenure into leasehold tenure (GRZ, 1995).

What Zambia needs to learn is how to provide a dynamic free enterprise marketing chain in the rural sector to replace the marketing boards which have been phased out (Meadly, 1990). This is necessary but it is likely to take some time to fill the existing vacuum in input provision and crop marketing which seems to have adverse effects on small-scale farmers. Herrera et al (1997) contend that agricultural reform requires the development of efficient and productive agricultural and a comprehensive institutional framework that ensures rights and security. They further argue that the creation of both public and private institutions requires the enhancement of the role of civil society, such as NGOs, in order to fill the institutional vacuum left by the reduced role of the state. It would seem that Zambia will need to promote further the role of NGOs as some progress is already underway to help small-scale farmers adjust to the new neo-liberal policies (Mitti, Drinkwater and Kalonge, 1997).

While the provision to convert communal land into leasehold is meant to confer secure land tenure rights to individuals, it might be further useful for Zambia to adopt the strategy that is being implemented in South Africa. Ntsebeza (1999) indicates that in the Eastern Cape, the government is attempting to balance the constitutional requirements of the needs of democratically elected representatives with those of hereditary traditional rulers. This is being done by taking care of land rights which belong to individuals and groups. Group rights are being vested in families but not in institutions such as the tribe.

The recognition of group land rights may be seen as a transitional stage aimed at incorporating traditional leaders especially in those areas where land markets are not yet developed. In this way, individualisation of agricultural land tenure will evolve gradually over time.

Alternatively, it could be helpful to consider the strategy in Uganda where all citizens owning land under customary tenure may acquire a certificate of customary ownership. These certificates may be leased, mortgaged and pledged where the customs of the community allow (Mwebaza, 1999). It may be argued, again, that this is a transitional step towards individualisation of land tenure in an evolutionary approach, which avoids alienating the traditionalists and other groups especially where land markets are not yet fully developed.

The purpose for such changes in agricultural land tenure is to create an enabling environment that can promote business approaches to farming by small-holders, who can contribute to agricultural growth, sustainable livelihoods and development.

Forest Management

Although the Zambian government recently formulated a forestry policy of jointly managing the forest estate with other stakeholders, such as individuals, local communities, NGOs and the private sector (GRZ, 1988), it is imperative that the country considers experiences of other countries in order to find a suitable model that can be implemented locally.

We have noted above, for instance, that in countries of Central and Eastern Europe, the forest estate has been privatised and individual small-holders will have to manage their own forests instead of depending on their governments. In India and Nepal, the strategies are rather different. In these countries, the forest departments are closely working with the forest user groups at the village level to manage the resources and engage in afforestation programmes to restore degraded forests.

Since Zambia has not yet implemented its joint forest management (JFM) policy, it will require making careful consideration as to what kind of institutions will be empowered to manage the forest estate at the local level. Are there existing indigenous institutions which can be assisted to undertake this responsibility or will new ones be created?

It is also worth noting that in Asia community forestry has begun to emphasise the significance of income generation for the individuals and communities involved in forest management. This is necessary in order to motivate individuals and raise living standards. Similar considerations will have to be made in Zambia so that the management of the forest estate by stakeholders contributes to the raising of living standards, sustainable livelihoods and the preservation of watersheds and biological diversity.

With respect to afforestation programmes, it has been argued that Zambia is lagging behind the trend in the neighbouring East African countries where ambitious rural or village afforestation programmes are taking place (Chidumayo, 1988). Furthermore, afforestation programmes in Zambia have performed below expectations not only because seedlings were distributed to growers free of charge, but also because no attention was paid or no follow-up measures were undertaken after distribution.

It is worth noting that tree planting in East Africa, and especially in Kenya, has taken off within the context of individualisation of tenure. With the decline of access to communal trees, farmers have opted to plant their own trees as boundaries and sources of building poles, fuelwood and for income generation (Scherr, 1997). Tree planting in Zambia may not take off significantly for a long time to come as long as small-scale farmers still have the option of obtaining building poles and fuelwood from the abundant communal forests in many parts of the country. However, planting of multi-purpose trees should be promoted in those regions such as Central, Copperbelt, Southern and Lusaka provinces where natural forests have disappeared due to clearance for agriculture and for charcoal (Chudumayo, 1988a; Kerkhof, 1990a).

The Role of Women in Forest Management and Agroforestry

Literature on the role of women in forest management and agroforestry in Zambia is not available. It is known, however, that

Zambian women in rural areas, like other rural women in the rest of Africa and other developing countries are dependent on natural forests for the extraction of forest resources such as fuelwood, grass for thatching, mushrooms, wild fruit and rodents. They also depend on forests as sources of herbal medicines.

In his study of village tree planting in Zambia, Chidumayo (1988) observed that in terms of gender participation, private woodlots were planted either by husbands or by husbands and their sons. Women did not participate in tree planting as much as men. This was so partly because since men wanted to plant the trees for sale, they discouraged women from participating so that they could also have total control over any income. Furthermore, since men were also more involved in firewood collection as the wood resources diminish, it was therefore advantageous for them to plant trees in order to reduce their firewood collection burden.

Similarly, Kerkhof (1990) found out in his study of an agroforestry project in southern Zambia that the proportion of women farmers participating in the meetings at which agroforestry information and seedlings were being disseminated, was only about 5%. This was a serious drawback as women do most of the agricultural work.

He reported that there were instances in which women pulled up Acacia albida seedlings which their husbands had planted, but mistook them to be weeds.

Although women in many rural areas of Zambia have access to communal forest resources, the impending introduction of JFM by the government should ensure that women are equally involved as stakeholders. More efforts should be made through female extension officers to reach a large number of women and girls.

Furthermore, the legal provision under the Lands Act of 1995 which allows small-scale farmers who occupy customary land to convert it to leasehold, should also apply to women. Women should not be left out in this gradual process of individualisation of agricultural land tenure. Women will need to be guaranteed security of tenure independent of their husbands, so that if they plant trees, such trees and the income resulting from their sales can

accrue to them. In this way, trees which women plant in
agroforestry projects can act as long-term savings (Chambers and
Leach, 1987).

Conclusion

This paper which is based on the review of literature, has
attempted to highlight trends which are taking place in developing
countries in terms of land reform, forest management and the role
of women in forest management and agroforestry, in order to
establish lessons for Zambia.

It has been shown that in Latin America and Asia, the winds of
neo-liberalism, since the collapse of the former Soviet Union have
ushered in policies of privatisation of land in free markets,
replacing the old policies of state control of land and input
provision. These changes emphasise the role of the private sector
and the need for secure individual tenure.

With respect to sub-Saharan Africa the paper has attempted to
show that apart from South Africa and Zimbabwe, where issues of
land redistribution, restitution and resettlement are on the agenda,
the debate is on how to harmonise customary tenure with modern
individual tenure, in order to create an enabling environment in
which small-scale farmers can have secure land rights and be
motivated to engage in business oriented agriculture.

It has also been shown that new approaches in forest
management have emerged in Asia and Central and Eastern
Europe. While India and Nepal have adopted related policies of
joint forest management and community forestry, in which local
communities are empowered to manage forests and restore
degraded ones through tree planting, the countries in Central and
Eastern Europe have privatised the forest estate. Small-holder
forest owners will have to manage their holdings probably in
collaboration with government and private extension efforts.

We have noted that although women have managed forests for
a long time, from which they extract fuelwood and other non-
timber forest products, their participation in agroforestry projects
is hindered by tradition and lack of secure land and tree tenure.

It is imperative that Zambia, which has embarked on policies of converting customary land tenure to leasehold tenure and introduced joint forest management, works out suitable models which do not only take into account external trends, but also consider and adopt the trends and models to suit local conditions. This will also require ensuring that small-scale farmers especially women, are empowered with secure title to land and the trees which they plant, so that both agricultural production and agroforestry programmes become sources for income generation, sustainable livelihood and environmental protection.

Acknowledgement

The review of literature upon which this paper is based, was made possible by a private sponsorship to undertake sabbatical leave at the University of Reading, by Mr Colin Trapnell, Bristol, England. He is a former government ecologist in northern Rhodesia (Zambia). However, the views expressed are entirely the responsibility of the author.

References

Binswanger, H.P. and Deininger. 1993, "South Africa Land Policy: The Legacy of History and Current Options", in *World Development*, vol.21, no.9, pp.1451-1475.

Bradley, P.N. 1991, *Woodfuel, Women and Woodlots*, vol.1, Macmillan Education Ltd.

Brokensha, D. and Riley, B. 1988, "Forest, Foraging, Fences and Fuel in a Marginal Area of Kenya", in Fortmann, L. and Bruce, J.W. (eds), *Whose Trees? Proprietary Dimensions of Forestry*, Westview Press, London, pp. 102-105.

Bruce, J.W. and Fortmann, L. 1988, "Why Land Tenure and Tree Tenure Matter: Some Fuel for Thought", in Fortmann, L. and Bruce, J.W. (eds), *Whose Trees? Proprietary Dimensions of Forestry*, West View Press, Boulder and London, pp. 1-9.

Bruce, J.W. 1993, "Do Indigenous Tenure Systems Constrain Agricultural Development ?", in Bassett, T.J. and Crummey, D.E.(eds),

Land in African Agrarian Systems, The University of Wisconsin Press, pp. 35-56.

Byron, N. 1997, "Income Generation Through Community Forestry", in Victor, M.(ed.), *Income Generation Through Community Forestry*, Proceedings of an International Seminar held in Bangkok, Thailand, October 18-20, Report No. 13, pp. 1-14.

Campell, J.Y.1995, "Foreword", in Roy, S.B.(ed), *Enabling Environment for Joint Forest Management*, Inter-India Publications, New Delhi.

Chambers, R. and Leach, M. 1987, "Trees to Meet Contingencies: Savings, and Security for the Rural Poor", in *Social Forestry Network*, ODE, Network Paper No. 59, October.

Chidumayo, E.N. 1988, "Village Tree Planting in Zambia: Problems and Prospects", in *Desertification Control Bulletin*, No. 17, pp. 22-26.

Chidumayo, E.N. 1988a, "Integration and Role of Planted Trees in a Bush-fallow Cultivation System in Central Zambia", in *Agroforestry Systems*, 7, pp. 63-76.

Chinene, V.R.N. et al, 1998, "A Comparison of Customary and Leasehold Tenure: Agriculture and Development in Zambia", in *Land Reform, Land Settlement and Cooperatives*, FAO, No. 2, pp. 89-99.

Choudhury, P.K.R. 1995, "Participatory Monitoring and Evaluation", in Roy, S.B. (ed), *Enabling Environment for Joint Forest Management*, Inter-India Publications, New Delhi, pp. 187-200.

Dankelman, I. and Davidson, J. 1988, *Women and Environment in the Third World: Alliance for the Future*. Earthscan Publications, London and IUCN.

Deville, P.L. 1999, *Harmonising Formal Law and Customary Land Rights in French-speaking Africa*, International Institute for Environment and Development (IIFD), Issue Paper No. 86, June.

Dobriyal, K. and Ballabh, V. 1995, "A Comparative Study of JFM in West Bengal and Gujarat", in Roy, S.B. (ed), *Enabling Environment for Joint Forest Management*, Inter-India Publications, New Delhi, pp. 213-230.

El-Lakany, H. 1997, "Preface", in *Issues and Opportunities in the Evolution of Private Forestry and Forestry Extension in Several Countries with Economics in Transition in Central and Eastern Europe*, FAO, Rome.

FAO and SIDA, 1991, *Restoring the Balance: Women and Forest Resources*, Rome and Sweden.

Faure, A. 1995, *Private Land Ownership in Rural Burkina Faso*, International Institute for Environment and Development (IIED), Issue Paper No. 59, October.

Fernandes, E.C.M. et al, 1989, "The Chagga Home Gardens: A Multi-storeyed Agroforestry Cropping System on Mount Kilimanjaro, Northern Tanzania", in Nair, P.K.R. (ed), *Agroforestry Systems in the Tropics*, Kluwer Academic Publishers and ICRAF, London, pp. 309-325.

Fortmann, L. and Bruce, J.W. 1988, "Tree Tenure", in Fortmann, L. and Bruce, J.W. (eds), *Whose Trees? Proprietary Dimensions of Forestry*, Westview Press, London, pp. 11-15.

Fortmann, L. and Bruce, J.W. 1988a, "The Gender Division of Tenure", in Fortmann, L. and Bruce, J.W. (eds), *Whose Trees? Proprietary Dimensions of Forestry*, Westview Press, London, pp. 235-239.

Fu, C. and Davis, J. 1998, "Land Reform in Rural China since the mid 1980s", in *Land Reform, Land Settlement and Cooperatives*, No.2, FAO, pp. 123-137.

Funnell, D.C. 1991, *Under the Shadow of Apartheid: Agrarian Transformation in Swaziland*, Avebury, Aldershot.

Gilmour, D.A. and Fisher, R.J. 1991, *Villagers, Forests and Foresters*, Sahayogi Press, Kathmandu.

GRZ, 1995, *The Lands Act, 1995*, Government Printer, Lusaka.

GRZ, 1998, *National Forestry Policy*, Ministry of Environment and Natural Resources, Government Printer, Lusaka, July.

Haque, T. and Montesi, L. 1996, "Tenurial Reforms and Agricultural Development in Vietnam", in *Land Reform, Land Settlement and Cooperatives*, FAO, pp. 67-77.

Herrera, A. et al, 1997, "Recent FAO Experiences in Land Reform and Tenure", in *Land Reform, Land Settlement and Cooperatives*, No. 1, FAO, pp. 53-64.

Hunt, S.M. et al, 1997, "Income Generation Through Community Forestry in Nepal", in Victor, M. (ed), *Income Generation Through Community Forestry, Proceedings*, Bangkok, Thailand, October 18-20, Report No. 13, pp. 63-80.

Hyden, G. 1986, "The Anomaly of the African Peasantry", in *Development and Change*, 17, (4), pp. 677-705.

Kay, C. 1998, "Latin America's Agrarian Reform: Lights and Shadows", in *Land Reform, Land Settlement and Cooperatives*, No. 2, FAO, pp. 9-31.

Kerkhof, P. 1990, "Lessons from Experience", in Foley, G. and Barnard, G. (eds), *Agroforestry in Africa: A Survey of Project Experience*, PANOS, pp. 3-10.

Kerkhof, P. 1990a, "Soil Conservation and Agroforestry Project Zambia", in Foley, G. and Barnard, G. (eds), *Agroforestry in Africa: A Survey of Project Experience,* PANOS, pp. 143-148.

Maharjan, M.R. 1997, "Income Generation Through Community Forestry: Case Studies from the Koshi Hills of Nepal", in Victor, M. (ed), *Income Generation Through Community Forestry, Proceedings,* Bangkok, Thailand, October 18-20, Report No.13, pp. 81-92.

Marghescu, T. and Anderson, J. 1997, "Overview", in *Issues and Opportunities in the Evolution of Private Forestry and Forestry Extension in Central and Eastern Europe,* FAO, Rome, pp. 3-44.

Meadley, J. 1990, "A More Significant Role for the Private Sector in Agricultural Development", in Speedy, A. (ed), *Developing World Agriculture,* Grosvenor Press International, pp. 14-19.

Mitti, G., Drinkwater, M., and Kalonge, S. 1997, "Experimenting with Agricultural Extension in Zambia: Care's Livingstone Food Security Project", in *AGREN,* ODI, Network Paper No. 77, July.

Mwebaza, R. 1999, *How to Integrate Statutory and Customary Tenure: The Uganda Case,* IIED, Issue Paper No. 83, June.

Nsamba-Gayiiya, E. 1999, *Implementing Land Tenure Reform in Uganda: A Complex Task Ahead,* IIED, Issue Paper No. 84, June.

Ntsebeza, L. 1999, *Land Tenure Reform in South Africa: An Example from the Eastern Cape,* International Institute for Environment and Development (IIED), Issue Paper No. 82, June.

Obi, C. 1988, "Women's Rights and Interests in Trees", in Fortmann, L. and Bruce, J.W. (eds) *Whose Trees? Proprietary Dimensions of Forestry,* Westview Press, London, pp. 240-242.

Palit, S. 1995, "JFM in India: Major Issues", in Roy, S.B. (ed), *Enabling Environment for Joint Forest Management,* Inter-India Publications, New Delhi, pp. 25-33.

Plateau, J.P. 1992, *Land Reform and Structural Adjustment in Sub-Saharan Africa: Controversies and Guidelines,* FAO, Rome.

Rocheleau, D.E. 1988, "Women, Trees and Tenure: Implications for Agroforestry", in Fortmann, L. and Bruce, J.W. (eds), *Whose Trees? Proprietary Dimensions of Forestry,* Westview Press, London, pp. 254-272.

Scherr, S.J. 1997, "Meeting Household Needs: Farmer Tree Growing Strategies in Western Kenya", in Arnold, J.E.M. and Dewees, P.A. (eds), *Farms, Trees and Farmers: Responses to Agricultural Intensification,* Earthscan, London, pp. 141-173.

Smyth, I. 1999, "Women, Industrialisation and the Environment in Indonesia", in Afshar, H. and Barrientos, S. (eds), *Women, Globalisation and Fragmentation in the Developing World*, Macmillan Press Ltd., pp. 131-149.

Speirs, M. and Marcussen, H.S. 1988, *Limits to Environmental Planning in a World of Structural Adjustment: The Case of Burkina Faso*, IIED, Issue Paper No. 75, April.

Ssembajjwe, G. 1997, "Joint Forest Management in Tanzania: Uganda Farmers Learn from the Duru-Haitemba Experience", in *Forest Action News*, Vol.1, No.8, p. 1 and 6, February.

Thiesenhusen, W.C. 1995, *Broken Promises: Agrarian Reform and the Latin American Campesino*, Westview Press, Oxford.

Thomson, J.T. 1988, "Participation, Local Organisation, Land and Tree Tenure: Future Directions for Sahelian Forestry", in Fortmann, L. and Bruce, J.W. (eds) *Whose Trees? Proprietary Dimensions of Forestry*, Westview Press, London, pp. 204-214.

Timberlake, L. 1988, *Africa in Crisis: The Causes, the Cures of Environmental Bankruptcy*, Earthscan Publications Ltd., London.

Turner, S.D. 1988, "Land and Trees in Lesotho", in Fortmann, L. and Bruce, J.W. (eds), *Whose Trees? Proprietary Dimensions of Forestry*, Westview Press, London, pp. 199-203.

Victor, M. (ed), 1997. *Income Generation Through Community Forestry*, Proceedings of an International Seminar held in Bangkok, Thailand, October 18-20, Report No. 13 (Foreword).

Von Blanckenburg, P. 1994, "Land Reform in Southern Africa: The Case of Zimbabwe", in *Land Reform, Land Settlement and Cooperatives*, FAO, pp. 5-14.

Warner, K. 1997, "Patterns of tree growing by farmers in Eastern Africa", in Arnold, J.E.M. and Dewees, P.A. (eds), *Farms, Trees and Farmers: Responses to Agricultural Intensification*, Earthscan, London, pp. 90-137.

21

Environmental Degradation in the Madhupur Tract

Mirza Mafizuddin and Mesbah-us-Saleheen

Madhupur tract occurring on the central part of Bangladesh occupies an area of about 4000 square kilometers of highlands between the Jammu and the old Brahmaputra floodplain (Figure 1). The tract has undergone upliftment in recent times and is composed of older alluvium of the ancient river (Morgan and McIntire 1959). It comprises an area of highly dissected, faulted and weathered soil cover on the surface. Because of the long exposure, the constituent materials in the tract have been oxidised and weathering and surface water erosion has led to the formation of fine textured alluvium on the surface area. Vast area of the tract, owing to highlands and fine texture has provided ample scope to have a sound vegetation cover stretching over an area of 0.11 million hectares. Innumerable minor species may be present but Sal (*Shorea robusta*) occupies a vast area in the tract. Due to prevailing population explosion in Bangladesh, people are encroaching the Madhupur tract to have extended agricultural land or land for settlement purpose. As a result of industrial establishment, settlement encroachment and forest clearing, the environment of Madhupur tract is at stake. An attempt has been made in this paper to assess the degradation of the environment in

Figure 1
Location of the Study Area

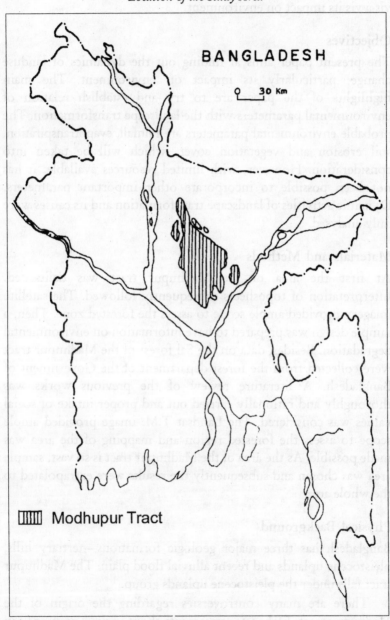

BANGLADESH

0 ___ 30 Km

||||| Modhupur Tract

the Madhupur tract in terms of depletion of forest resource vis-a-vis its impact on environment.

Objectives

The present paper aims at finding out the dynamics of landuse change, particularly its impact on environment. The main highlights of the paper are to try and establish relation of environmental parameters with the landscape transformation. The probable environmental parameters are rainfall, evapotranspiration, soil erosion and vegetation cover, which will be taken into consideration. However, with limited resources available, it has not been possible to incorporate other important parameters. Dynamic activities of landscape transformation and its causes were duly analysed.

Materials and Methods

At first the area of the Madhupur tract was delineated. Interpretation of toposheets subsequently followed. The satellite imagery provided ample scope to assess the forested zone. Then, a sample design was prepared to have information on environmental degradation. Besides, data on the Sal forest of the Madhupur tract were collected from the forest department of the Government of Bangladesh. A literature review of the previous works was thoroughly and rationally carried out and proper intake of social values was considered. The Landsat T.M. image provided ample scope to assess the forested region and mapping of the area was made possible. As the area of the Madhupur tract is so vast, sample area was chosen and subsequently the results were extrapolated to the whole area.

Physical Background

Bangladesh has three major geologic formations——tertiary hills, pleistocene uplands and recent alluvial flood plain. The Madhupur tract falls under the pleistocene uplands group.

There are many controversies regarding the origin of the Madhupur tract. Elaborate assessments about various opinions can be had from the writings of Moniruzzaman Mia and L. Bazlee (1968) and Bazlee (1967). According to the views of different authors, it was suggested that the Madhupur tract had been modified in recent times by the Ganges and the Brahmaputra rivers (Fergusson 1963, Hirst 1926, La-Touche, 1919). Rizvi (1957), Morgan and McIntire (1959) recognised the tract as a pleistocene terrace. They observed its similarity with Gulf Coast terraces and speculated its origin as floodplains of earlier Ganges-Brahmaputra river system. The Madhupur clay, as proposed by Khan (1953) has no fossil evidence and appears to be the terrestrial equivalent of the St. Martin's limestone formation and hence may be treated as a pleistocene formation (Khan 1991).

Lithology

Madhupur tract is located in the central part of the Bengal basin. This tract has undergone changes through time and traces can be marked on it in the present-day surface. Based on the available information it can be said that the present Madhupur tract got its shape not earlier than the Pleistocene age. Prior to that, surface geology of the tract was similar to elsewhere in Bengal basin to the extent of Cretaceous period (Guha 1978, Khan 1975 and Khan 1978).

The tract had been subjected to marine transgression effect during late Cretaceous, Eocene and between Miocene and Pliocene; it was subsequently followed by deltaic sedimentation. Towards the end of the Miocene a distinct erosion phase took place. The present Madhupur tract was raised during the Pleistocene period when the fifth Himalayan orogenic movement took place.

Geomorphology of the Madhupur Tract

Madhupur tract is unique in terms of its topography, drainage, relief and age factors of soil formation and is quite different from the adjoining floodplains. The tract shows elevation difference in different parts especially the western side; this elevation to the tune of 100 feet can be observed. The relief is extremely variable and fault scarp occurs on the western part (Morgan and McIntire 1959).

Elsewhere in the tract there is little or no break of slopes and in the eastern and southern part, it slopes down gradually beneath the overlapping recent sediments.

There are some isolated patches of the tract, which have been created by vigorous river action. Most of the terrace summit was flat, or undulating and rolling in topography. Eastward slope is 0.70 m/km, which indicates considerable eastward tilt of the tract. High terraces have leveled summit, locally called as *Chala*, and the narrow winding valley created as result of dissection is called *Byde*. The greater part of the tract lies above flood level. Because of dissection of the terrace, two types of valleys are formed. One is usually deep and narrow and the other is shallow and usually broad. The deep valleys are occupied with dark heavy clays and narrow valleys have light gray porous silt-loams to silty clay. The shallow valleys have no buried floors. They form a close dendritic type of drainage pattern. The terraces are broadly dissected terrace and closely dissected terrace. Some saddle and spur are formed due to this dissection process. Some of the rivers, inspite of progressive upliftment, maintain their course and as a result antecedent character is formed, for example the Bansi river.

High intensity of rainfall causes various processes of water erosion and as a result damaging effects are pronounced. Sheetwash is seen to have created the bare ground in the tract. The development of rill is related to seasonality of rainfall in the Madhupur tract. Erosion by seepage and through flow is not less important in the tract. Sample study of soil erosion reveals the existence of spatial variation of soil erosion phenomena (Mafizuddin and Alam 1983) and it has generated due to differential rates of geomorphic processes, relief and vegetation cover.

Climate

Like all other parts of Bangladesh, the Madhupur tract has characteristically high temperature, high radiation and a very definite seasonal distribution of rainfall. With a strong seasonal variability from year to year, a complete dry season of 4-7 months is followed by only rainy period. Throughout the wet season, rainfall is concentrated in a few days with frequent dry spells and at the initial and final stage rainfall intensity reaches high value, therefore the climate is mild and frost in unknown in the Madhupur tract. In terms of ground climatic conditions, the existence of micro-climate variation in the tract reflects the difference of landforms, surface elevation, drainage and vegetation cover within the tract (Mafizuddin and Alam 1982).

Soil

The soil belonging to the Madhupur tract originated from the typical Pleistocene alluvium and is specially known to the penologist as Madhupur clay. It is compact clay having poor drainage conditions on the terrace summits. The typical clays occupy the dissected areas called as *Byde*. However, on the terrace's surface clayloam or silty clay-loam soil persists. The soil colour ranges from yellowish to brown to reddish-brown. In the well-drained areas yellowish-red is found and in poorly drained areas olive brown or grayish-brown are found. The soils are less porous having minimum organic matter content, and weathered mineral constituents are leached out of it.

Friable clay-loam with organic rich terrace soils provide ample scope for luxuriant growth of vegetation. Sheetwash, rillwash and gullies and spontaneous growth of natural vegetation. So, in this degraded forest, only thorny bushes and (scrubs sometimes with their exposed roots due to surface water erosion) remain to be seen.

The subsoil is deeply and strongly weathered and of strongly molted clay which is widespread with their uniform characteristics. Iron and manganese concentrations are the typical feature of the soils of the shallow valleys and the somewhat poorly drained areas, and calcareous nodules are abundant in the soils of closely dissected areas. The closely dissected terrace areas are prone to surface and subsurface water erosion, thereby creating denuded topography where soil development is poor.

Figure 2
Distribution of Major Forest Areas of the Madhupur Tract

FOREST AREA
1915

Results and Discussion

The study of topographic sheets of the scale 1:50,000 provided important scope to identify the nature of forest cover in the area. Since the area has undergone transformation, the 1915 (Figure 2) and 1999 (Figure 3) maps gave a clear and supportive picture of the study area in terms of the depletion of resources.

The sample areas provided immense scope to assess the environmental degradation. First, the Chandura mouza, having an area of 340 hectares of forested lands in 1952, declined sharply. The loss of forest land to the tune of 25% bore a severe impact around the area. The 1952 photograph indicated the forest cover of the

Figure 3
Distribution of Forest Areas of the Madhupur Tract

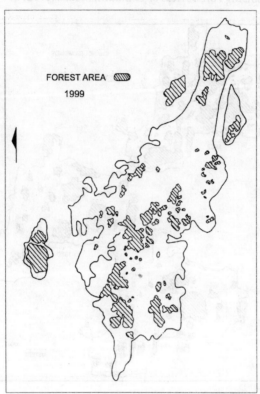

Chandura sample area to be 340 hectares but subsequently in 1980 (Figure 4), it reduced to 110 hectares, in 1989 (Figure 4) to 105 hectares, in 1992 (Figure 5) to 99 hectares and in 1999 to 90 hectares (Figure 6) hectares respectively. The forest cover of the sample area has been reduced to 25% of the total area during the last 40 years.

Due to population increase in the sample area, demand for new lands for cultivation has sharply risen. Settlement encroachment and demand for fuelwood force people to grab the forested area, and this has so happened due to internal migration of people from around the areas. The depletion of forest resources was evidenced

Figure 4
Map Showing Forest Areas of Chandura Mouza (1952, 1980, 1989)

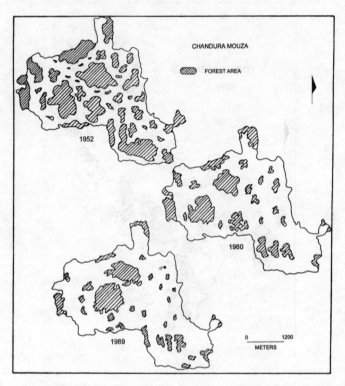

from the analysis of the organic matter content of the soils of the area. It showed a significant difference in organic matter in forested (1.48%) and deforested area (being 0.53%) and partially forested area (1.085%) (Sady and Mafizuddin, 1992). This paves the way for soil erosion, as effects of precipitation and surface runoff will have accelerated rate of natural removal. This changing pattern of ecosystem will be detrimental for the living conditions of flora and fauna. Above all, the bio-geochemical cycle will be disturbed in the area. Therefore, spatial variation in the terrain in terms of organic matter content may be attributed to the ecological disturbance in the area.

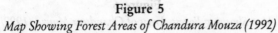

Figure 5

Map Showing Forest Areas of Chandura Mouza (1992)

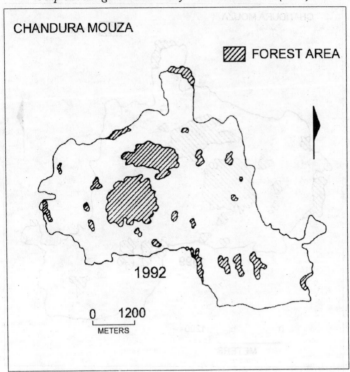

Deforestation also resulted in extinction of various wildlife in the area. Presently, no traces of monkeys and panthers can be marked which were abundant in the area some 50 years back. Soil moisture content is declining, and then again due to surface cover biomass production is going down.

Second sample area in Gazipur, particularly centering Salna area, revealed further deteriorated situation. The Bhawal Garh is famous in this area having a forest cover of Gazari, Sal, Teak and miscellaneous minor trees. The area being located within the closed proximity of Dhaka city, industrial expansion and population settlement have engulfed the area to some extent. The result is large-scale decline of forest area. The forest area was 20,900 ha in 1980 (Figure 7), which sharply declined to 15,300 ha in 1989

Figure 6
Map Showing Forest Areas of Chandura Mouza (1999)

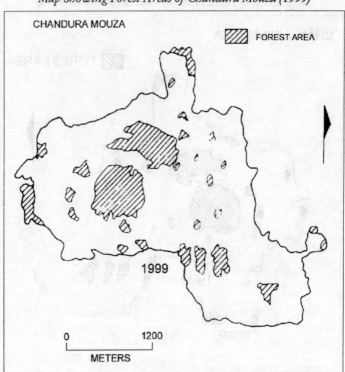

(Figure 8), 13,600 ha in 1922 (Figure 9) and 11,250 ha in 1999 respectively (Figure 10). The decline of the forest cover was to the tune of 45% of the total forested area.

The field mapping indicated decrease in forest and paucity of land ever for agriculture. The large-scale transformation is due to the expansion of industries in the area. The percentage of the organic matter is also similarly reducing at an alarming rate between the vegetated and non-vegetated regions. The marked change in Gazipur is that the forested area is located around the settlement area and is sparsely distributed. The area is dissected by valleys and some fallow lands remain in between. The forested area can easily be squeezed out by simple intrusion from all sides

Figure 7
Map Showing Forest Areas of Gazipur (1980) .

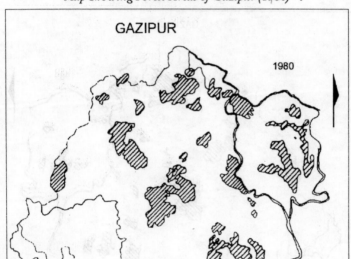

(Chowdhury and Ahmed 1992). The declining forest cover in Gazipur is similar to Chandura but here the population pressure is tremendously high. Gazipur was once a habitat for wildlife of different species, but now their numbers have reduced drastically. Biological weathering is being hampered and existing poor soil cannot be improved unless organic matter content increases, structural development is well defined and soil drainage condition is turned into well-drained status.

Third sample area is around Madhupur proper, the Arankhala mouza. The landscape transformation from 1915 to 1999 reflected a decline of forest cover of about 370 hectares from a total of 850 hectares. The vast majority was in the forested area in 1915 being 650 ha. During the period of 1915-16, the area under forest was

Figure 8
Map Showing Forest Areas of Gazipur (1989)

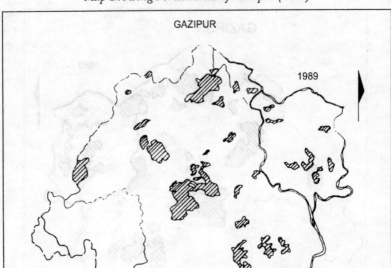

about 75% of the total area. The rate of decline is also alarming being 363 ha in 1980 (Figure 11), 327 ha in 1989 (Figure 12), 305 ha in 1992 (Figure 13) and 275 ha in 1999 (Figure 14) respectively. The area cleared from the forest is now being used for agricultural and settlement purpose, and few other areas are remaining as fallow lands. During the last 84 years, the forest has been reduced to 75 ha only. Resultant effects arising from this have been one of the major reason for degrading the environmental conditions of the area. The settlement encroachment, extension of agricultural lands and increasing accessibility of the forest area for private thoroughfare are the main causes for this deterioration of the environment. Forcible occupation of the area by the poor migratory people has also reduced the forested area considerably. The timber smugglers

Figure 9
Map Showing Forest Areas of Gazipur (1992)

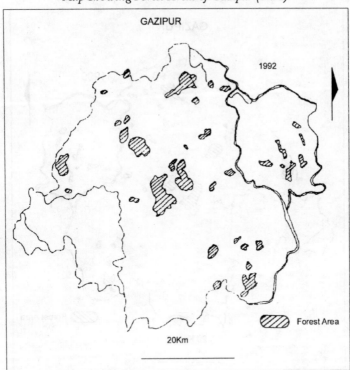

have further added to the rising problem of forest decline. Because the forest lands are the public property, trespassers cut and take away trees in a organised manner. The corrupted employees of the forest department also help the smugglers in this venture (Alim 1996).

Conclusion

Stability of environment rests on the availability of vegetation cover on the earth's surface. Economic interventions in forests are bound to change the quality and quantity of energy, material and information in the ecosystem. Instability leads to destruction of resource potential in an area. Due to prevailing instability the biological diversity is also affected.

Figure 10
Map Showing Forest Areas of Gazipur (1999)

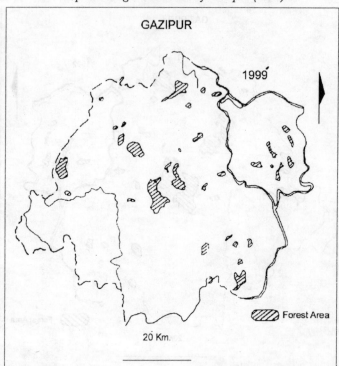

Anthropogenic disturbances play a key role in removing the balance of the environment ecosystem. Deforestation of the forest, soil erosion and lack of soil moisture are the important phenomena that are caused by human endeavor. Even the developed countries are not spared from this menacing problem. In USA, one million sq km area is threatened by erosion and in Russia 500,000 sq km of land has been eroded by water. Environmental degradation is imminent if the pronounced activities of deforestation are continued. It causes the break-up of a protective grass cover and intensive cultivation of areas. It accentuates a driving force, which influences: (i) population growth, and (ii) inequitable social conditions.

Figure 11
Map Showing Forest Areas of Arankhola (1980)

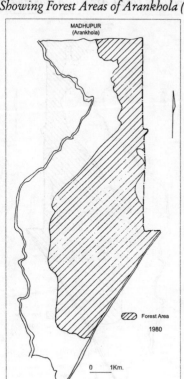

The extreme deforestation in the Madhupur tract may have several adverse ecological effects, such as:

1. Disruption of hydrological regime;
2. Soil moisture retention capacity decline;
3. Increase of sediment load of the stream plying in the tract because of exposure of surface and thereby accelerating soil erosion;
4. Changes in the surface energy budget;
5. Alteration of bio-geochemical cycles;
6. Disbalance in bio-diversity of species;
7. A decrease in the ecological complexity of ecosystem.

Figure 12
Map Showing Forest Areas of Arankhola (1989)

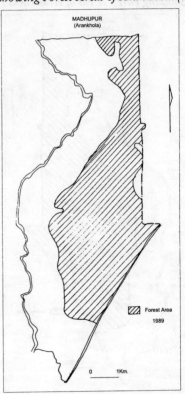

Economic development may be affected as consumptive use of forest by people will be hindered, and commercial wood supply will be restricted and fuelwood shortage will be acute. Therefore, to restore the environment in the tract, certain steps should be taken in such a way that the ecosystem's productivity increases substantially. A scheme for regeneration of vegetation should be undertaken to overcome the deficiency of ecological diversity. Soil erosion by human interference should be kept minimum so that nutrient loss of the surface is checked. Tree cutting, smuggling and unauthorised trespassing should be restricted through enactment of law.

Figure 13
Map Showing Forest Areas of Arankhola (1992)

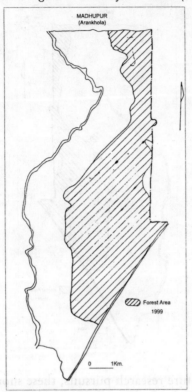

Micro-climatic variation should not be left to vary, in other words depletion of forest resource in no way should be allowed to continue. Biomass production should be increased and nutrient recharge should be kept open through complex interaction of organic activity. Effect should be concentrated to limit the alteration of bio-geochemical cycles. Although the present study reflected a pattern in sample area, the problems and prospects of the study can be extrapolated to the whole of Madhupur tract.

Because of time constraint, the study could not incorporate large number of variables. Ideally, a proper and pragmatic sample design should be followed which involves stratified random sample

Figure 14

Map Showing Forest Areas of Arankhola (1999)

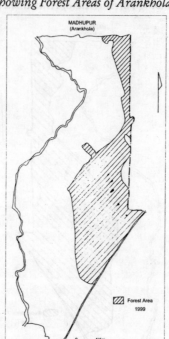

procedures. In future research pursuits, these should be given due consideration.

References

Alim, M.A. 1996: The depletion of Sal forest in the Madhupur Tract: A case study of Arankhola and Gachhbari mouza. Unpublished M.Sc. Research Report. Department of Geography and Environment, Jahangirnagar University, Savar, Dhaka.

Bazlee, L.N. 1967: Geomorphological study of the southern part of the Madhupur Terrace. Unpublished M.Sc. Thesis. Department of Geography, University of Dhaka, Dhaka.

Chowdhury, M.H. and Raquibuddin, A. 1992: Multi-temporal analysis of land change using different satellite platforms and GIS integration. *Journal of the Bangladesh National Geographical Association*, Vol. 20, No. 1 and 2.

Furgussion, J. 1863: Delta of the Ganges: Recent Changes. *Journal of Geology*, Vol. XIX, London.

Guha, J.K. 1978: Tectonic framework and oil and gas prospects of Bangladesh. Proceedings of the Fourth Bangladesh Geological Conference, Dhaka.

Hirst, F.C. 1916: *Report on the Nadia Rivers*, Calcutta.

Khan, F.H. 1962: Clay deposits of East Pakistan, CENTO Symposium of Industrial Rocks and Minerals held in Lahore, Pakistan, December 1962.

Khan, F.H. 1978: Exploration of mineral deposits in Bangladesh. Proceedings of the Fourth Bangladesh Geological Conference, Dhaka.

Khan, F.H. 1991: *Geology of Bangladesh*. Wiley Eastern Limited, New Delhi.

Khan, M.A. 1975: Oil and gas prospects in Bangladesh. Proceedings of the Third Bangladesh Geological Conference, Dhaka.

La Taouche, T.H.D. 1919: Relics of the great ice age on the plains of Northern India. Reprinted in report on the Hoogly river and its headwater. *The Bangal Secretariat Book*, Vol. 1, Calcutta.

Mafizuddin, M. and Alam, M.S. 1982: A study on micro-climate variation in the Madhupur Tract. Jahangirnagar Review, Vol. 6 Part II, Dhaka.

Mafizuddin, M. and Alam, M.S. 1983: Watertable aggregate test as a measure of spatial variation of erodibility: A case study of the Madhupur Tract. *Oriental Geographer*, Vol. 27, No. 1 and 2, Dhaka.

Mia, M.M. and Bazlee, L.N. 1968: Some aspects of the geomorphology of the Madhupur Tract. *Oriental Geographer*, Vol. 12, No. 1, Dhaka.

Morgan, J.P. and McIntire, W.G. 1959: Quarternary geology of the Bengal Basin, East Pakistan and India. *Bulletin of the Geological Society of America*, Vol. 70.

Rizvi, A.I.H. 1957: Pleistocene terraces of the lower Ganges valley. *Oriental Geographer*, Vol. 1, Dhaka.

Sady, M.A.B.S.A. and Mafizuddin, M. 1992: Deforestation and its environmental impacts: A case study in Chandura Mouza. Seventh Bangladesh National Geographical Conference, Jahangirnagar University, Savar, Dhaka.

22

Some Aspects of Maximum One-day Rainfall Distribution Over West Bengal

B.N. Mandal, R.B. Sangam and A.K. Kulkarni

Optimum utilisation of available water resources of a region in our country has become the need of the hour owing to population explosion and rapid strides in agriculture and industrialisation. For the optimum development of water resources of a region, some basic knowledge about rainfall over that region is essential. As for example, information about maximum rainfall of different probabilities is a pre-requisite for planning and designing of medium and minor hydraulic structures. Keeping this in view, the following rainfall studies were carried out for the West Bengal state using the available long-period rainfall data of stations in the region:

(i) Preparation of a generalised chart showing the highest observed one-day point rainfall,

(ii) Estimation and preparation of generalised charts of maximum one-day rainfall of 2 to 100 years,

(iii) Estimation of one day probable maximum point rainfall (PMP) using latest statistical techniques, and

(iv) Estimation of maximum 1, 2, 3...24 hours rainfall for return periods of 2 to 25 years using 18-years (1948-1965) autographic rainfall records at Calcutta (Dum Dum).

A brief resume of each of the above rainfall topics over the West Bengal state has been given in the succeeding sections of this paper so that the information may be useful to various agencies dealing with planning and designing of water resources of this region.

Rainfall Characteristics of West Bengal

The state of West Bengal consists of 17 districts with an area of 88752 km² (India, 1996). This state has been divided in two homogenous meteorological sub-divisions of: (i) The Gangetic West Bengal, and (ii) The Sub-Himalayan West Bengal. Generally the southwest monsoon sets in over these two sub-divisions in the first week of June and withdraws by about the second week of October (Dhar et al., 1985). Both these sub-divisions experience tropical disturbances like cyclonic storms and depressions from the Bay of Bengal quite frequently during the monsoon months of June to September and occasionally during pre-monsoon (March to May) and post-monsoon months (October and November). Agriculture plays a pivotal role in the State's income and nearly three out of four persons in the State are directly or indirectly involved in agriculture (India, 1996). Seasonal and annual rainfall characteristics of the two meteorological sub-divisions in the state have been worked out using rainfall data of stations in the region for 80-year period (1891-1970) and the same are given in Table 1.

Table 1
Seasonal and Annual Rainfall (cm) over the Two Sub-divisions in West Bengal

Sr. No.	Name of the sub-division	Jan-Feb	% of A	Mar-May	% of A	June-Sept	% of A	Oct-Dec	% of A	Annual
1.	Sub-Himalayan West Bengal	2.8	1.0	42.5	15.2	219.1	78.3	15.5	5.5	279.8
2.	Gangetic West Bengal	3.6	2.5	17.4	12.3	107.8	76.0	13.1	9.2	141.9

(A = Annual rainfall)

It is seen from the above table that the mean annual rainfall over the Sub-Himalayan West Bengal is almost double than that over the Gangetic West Bengal. The southwest monsoon contributes about 78% of the annual rainfall over Sub-Himalayan West Bengal while it is about 76% of the annual over the Gangetic West Bengal. Pre-monsoon season (March-May) records about 15% and 12% of the annual respectively over these two sub-divisions in the State.

Highest Observed One-day Point Rainfall

Information about the highest observed one-day rainfall over a region is required by the design engineers and hydrologists for planning hydraulic structures of medium and minor nature. Keeping this in view, the highest one-day rainfall for 23 observatory stations uniformly distributed over the region were picked out from their monthly records from the "Climatological Tables of Observatories in India" (IMD, 1999). In these tables heaviest 24-hour rainfall are available from the starting of the observatory to the end of 1980. In addition to these 23 stations, a few selective stations' heaviest one-day rainfall were also collected from other sources in order to prepare a generalised chart. Highest one-day rainfall of each of these stations were plotted on a base map of the region, isohyets were then drawn and a generalised chart was made (Figure 1). From this figure it is seen that the highest one-day rainfall over the region can be of the order of 20 to 50 cm. The highest one-day rainfall at Darjeeling on 25 September, 1899 was found to be 49.3 cm.

If the highest ever recorded 24-hour rainfall by an individual station considering its records up to present time is taken into account, then a station called Hashimara in the Cooch-Behar district of north Bengal has recorded an unprecedented rainfall of 99.6 cm during 0530 IST of 20 July to 0530 IST of 21 July 1993. However, according to the standard practice of measuring rainfall at 0830 hours, a smaller value of 92.8 cm was recorded at this station (Srivastava and Mandal, 1995). Other notable one-day heaviest rainfall in recent times were 57 cm at Malda on 28 September 1995, and 61 cm at Sandheads on 15 May, 1995.

Figure 1
Generalised Chart of Heaviest One-day Rainfall over West Bengal

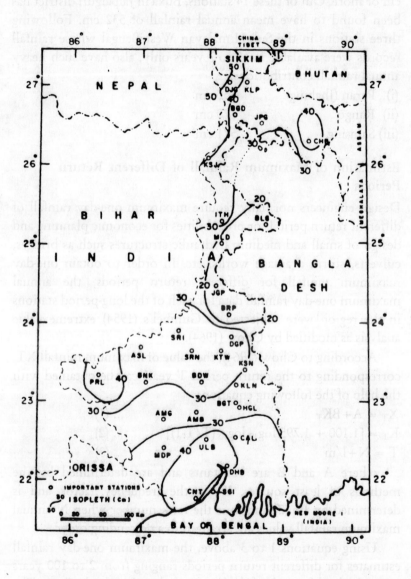

Dhar et al. (1978) have reported that in India there are 14 heavy rainfall stations which have recorded mean annual rainfall of 500 cm or more. Out of these 14 stations, Buxa in Jalpaiguri district has been found to have mean annual rainfall of 532 cm. Following three stations in the Sub-Himalayan West Bengal whose rainfall records were available for 5 to 7 years only, also have such heavy annual rainfall contributions:

(i) Paran (Jhalung)   650 cm
(ii) Rango   670 cm
(iii) Samsing   571 cm

Estimation of Maximum Rainfall of Different Return Periods

Design engineers normally require maximum one-day rainfall of different return periods or probabilities for economic planning and design of small and medium hydraulic structures such as bridges, culverts, storm drainage works, etc. In order to obtain one-day maximum rainfall for different return periods, the annual maximum one-day rainfall data for each of the long-period stations in the region were subjected to Gumbel's (1954) extreme value analysis as modified by Chow (1964).

According to Chow (1964), the value of maximum rainfall XT, corresponding to the return period T years can be obtained with the help of the following equations:

$$X_T = A + BK_T \qquad \qquad \text{... (1)}$$
$$K_T = [1.100 + 1.795 \log_{10} \log_{10} (T/T\text{-}1)] \qquad \text{... (2)}$$
$$T = N + l/m \qquad \qquad \text{... (3)}$$

where A and B are constants and are determined by the methods of least squares, K_T is the frequency factor and is determined by Eq.(2) and m is the rank number when N annual maximum rainfall values are arranged in a descending order.

Using equations 1 to 3 above, the maximum one-day rainfall estimates for different return periods ranging from 2 to 100 years were obtained for each of the long-period stations in the State. The maximum rainfall values for low return periods (i.e., 10 years) were converted into partial duration series by the application of relevant conversion factors obtained by Dhar and Kulkarni (1970).

One-day Generalised Rainfall Charts for Different Return Periods

The one-day maximum rainfall values for 2, 5, 10, 25, 50 and 100-year return periods for each of the stations were plotted separately on base maps of the region and isopleths of rainfall were drawn at suitable intervals and generalised charts were prepared. The generalised charts of 10 and 100-year return periods are shown in Figures 2 and 3.

From an examination of these charts it is seen that for 10-year return period maximum one-day rainfall over the region ranges from 15 to 30 cm while for 100-year return period these values vary from 25 to 45 cm. Estimates of rainfall values for other return periods over the region were found to vary as given in Table 2 below.

Table 2

Estimates of Maximum One-day Rainfall for Return Periods of 2, 5, 25 and 50 Years Over West Bengal

Return periods of	Range of maximum one-day rainfall (cm) over West Bengal
2-year	10-15
5-year	15-25
25-year	20-30
50-year	25-35

Probable Maximum Point Rainfall (PMP)

Probable maximum point rainfall (PMP) is defined as the highest or the extreme rainfall which the nature can produce over a given point or a specified area in a given duration of time. PMP estimates are used for the design of those hydraulic structures in whose case no risks due to their failure can ever be taken.

The estimates of PMP can be obtained by physical and statistical methods (WMO, 1973). The application of physical

Figure 2
Generalised Chart of 10-year One-day Maximum Rainfall Over West Bengal

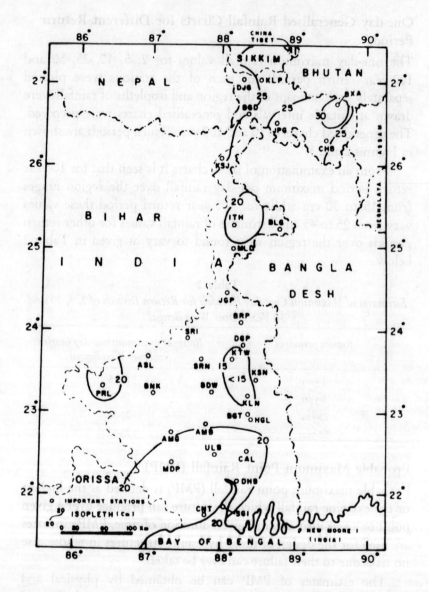

Figure 3
Generalised Chart of 100-year One-day Maximum Rainfall Over West Bengal

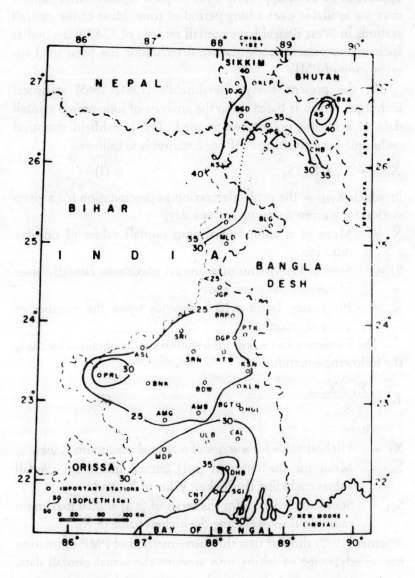

method which involves moisture maximisation, is doubtful in tropical regions (Dhar, 1972). In view of this, statistical technique appears to be more appropriate for tropical regions where rainfall data are available over a long period of time. Most of the rainfall stations in West Bengal have rainfall records of 70-80 years and as such in the present study statistical technique has been used for estimation of PMP.

In the present study, Hershfield's (1961, 1965) statistical technique which is based upon the analysis of long-period rainfall data of a station or an area, is used. The Hershfield statistical technique for estimating PMP for a station is as follows:

$$X_{PMP} = \overline{X}_n + K_m. S_n \qquad \qquad ... (1)$$

in which X_{PMP} = the probable maximum precipitation for a given station for a given duration, say one day,

$\overline{X}_n$ =   Mean of n annual maximum rainfall values of one-day duration,

S_n =   Standard deviation of n annual maximum rainfall values of one-day duration, and

K_m =   Frequency factor which depends upon the number of years of record.

The frequency factor km for a station can be obtained by using the following equation:

$$K_m = \frac{X_1 - \overline{X}_{n-1}}{S_{n-1}} \qquad \qquad ... (2)$$

where,

X_1 =   Highest value for a series of n annual maximum values,

$\overline{X}_{n-1}$ =   Mean for the series of (n-1) annual maximum rainfall values excluding the highest value of X_1, and

S_{n-1} =   Standard deviation for the series of (n-1) annual maximum rainfall values excluding the highest value of X_1.

Wiesner (1970) showed that the above method of PMP estimation has an advantage of taking into account the actual rainfall data, expressing it in terms of statistical parameters and is easy to use.

Using the statistical technique given above, estimates of PMP for all long-period stations in West Bengal were made for one-day duration and these values were then plotted on a suitable base map of the region, isopleths were drawn at suitable intervals and a generalised PMP chart for the region was prepared (Figure 4) (IITM, 1989). It is seen from this figure that estimates of one-day PMP over the study region range from 50 to 80 cm.

Short Duration Rainfall Analysis at Calcutta (Dum Dum)

Autographic rainfall records at Calcutta (Dum Dum) for a period of 18 years (1948-1965) has been analyzed. In this study, clock hour rainfall has been taken as the unit of study (Mandal et al., 1985). The observed maximum rain amounts recorded during each of the durations of 1, 2, 3, 6, 12, 18 and 24 hours were picked up for each year of the 18-year period from a scrutiny of the hourly data. Using Gumbel (1954) technique, maximum rainfall was worked out for different durations of 1, 2, 3,...24 hours for return periods of 2 to 25 years and the same are given in Table 3 below.

Table 3
Maximum Rain Amounts (in cm) for Different Durations and Return Periods

Duration (hours)	2-year	5-year	10-year	15-year	20-year	25-year
1	4.67	5.77	6.48	6.93	7.24	7.44
2	6.10	8.00	9.40	10.16	10.67	11.15
3	6.86	8.97	10.29	11.13	11.58	11.99
6	8.13	10.19	11.94	12.75	13.46	13.97
12	9.40	12.65	14.73	16.00	16.89	17.53
18	10.29	14.22	16.89	18.29	19.30	20.15
24	10.97	15.24	17.68	20.12	20.73	21.95

It is seen from the above table that estimates of 6 hourly rainfall at this station can vary from about 8.1 to 14 cm for 2 to 25 years return periods while for 12 hours estimated values can be of the order of 9.4 to 17.5 cm. From the scrutiny of hourly rainfall records of this station, 25 rainspells were selected which

Figure 4
Generalised Chart of One-day Probable Maximum Precipitation (PMP)
Over West Bengal

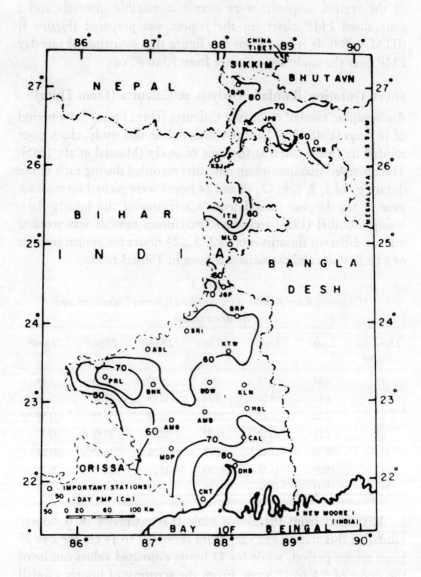

experienced 7.6 cm or more rainfall during 24 hours or more. The maximum rainfall recorded during different durations of 1, 2, 3, 6, 12, 18, 24 hours, etc., in each of these 25 rainspells were extracted and percentage ratios of maximum rainfall recorded in each of these durations, to that recorded during 24 and 48 hours were then worked out for each rainspell. The mean percentage ratios were then worked out for different durations and the same are given in Table 4 below.

Table 4

Mean Percentage Ratios of Different Durations to 24-hour and 48-hour Maximum Rainfall

	Duration (in hours) of							
	1	2	3	6	12	18	24	36
Mean % ratios to 24-hr rainfall	27	39	46	58	75	87	—	—
Mean % ratios to 48-hr rainfall	17	25	30	39	50	59	71	91

From the above table it is seen that in 6-hour duration about 58% of 24-hour rainfall can be recorded at this station while in 12-hour period about 50% of 48-hour rainfall can be recorded. Such autographic rainfall analysis is useful for the design of local drainage works, road culverts and railway bridges and also for estimating peak discharge for small basins.

Summary and Conclusions

From the foregoing analysis the following facts emerged from this study:

(1) Southwest monsoon (June-September) contributes about 76% of the annual rainfall over the Gangetic West Bengal and it is about 78% of the annual rainfall over the Sub-Himalayan West Bengal. Mean annual rainfall over the Sub-Himalayan West Bengal is almost double than that over the Gangetic West Bengal,

(2) Based on 23 observatory stations' data up to 1980 it is seen that the highest ever recorded one-day rainfall in the state can be of the order of 20 to 50 cm,

(3) Estimates of maximum one-day rainfall for 10-year return period can range from 15 to 30 cm while in 100-year period such estimated values are found to vary from 25 to 45 cm,

(4) The one-day PMP values for stations in the region are found to range from 50 to 80 cm, and

(5) Estimates of 6-hourly rainfall at Calcutta (Dum Dum) by considering autographic rainfall records of 18-year period (1948-1965) have been found to range from 8.1 to 14 cm for 2 to 25 years return periods while for 12 hours estimated values are found to vary from 9.4 to 17.5 cm.

Acknowledgements

Authors are grateful to Dr. G.B. Pant, Director of this Institute for his keen interest in hydrometeorological studies and to Dr. K. Rupa Kumar, Deputy Director for giving encouragement.

Annexure I

Name of the Rainfall Stations whose Abbreviations are put on Figs. 1 to 4

Abbreviation	Station Name	Abbreviation	Station Name
DJG	Darjeeling	BNK	Bankura
KLP	Kalimpong	SRN	Sriniketan
BGD	Baghdogra	BDW	Burdwan
JPG	Jalpaiguri	KSN	Krishnanagar
CBH	Coochbehar	KLN	Kalna
BXA	Buxa	BGT	Bagati
KST	Kishanganj	HGL	Hooghly
ITH	Itahar	AMB	Arambagh
BLG	Balurghat	ULB	Uluberia
MLD	Malda	MDP	Midnapore
JGP	Jangipur	DHB	Diamond Harbour
BRP	Berhampore	CNT	Contai
KTW	Katwa	SGI	Sagar Island
SRI	Suri	CAL	Calcutta (Alipore)
ASL	Asansol	DGP	Durgapur
PRL	Purulia		

References

Chow, V.T. (1964): Statistical and probability analysis of hydrologic data, Section 8.1, *Handbook of Applied Hydrology*, McGraw Hill Book Co., New York.

Climatological Tables of Observatories in India, 5th edition, (1999), India Meteorological Department (IMD) publication.

Dhar, O.N. and Kulkarni, A.K. (1970): Estimation of maximum one-day rainfall for different return periods in Uttar Pradesh, *Indian Journal of Met. Geophy.*, Vol.21, No.2.

Dhar, O.N. (1972): Chapter 6 on "Storm Transposition" in the *Manual of Hyrometeorology* published by India Meteorological Department.

Dhar, O.N., Mandal, B.N. and Ghose, G.C. (1978): Heavy rainfall stations of India: A brief appraisal, *Indian Journal of Power and River Valley Development*, Vol.28, No.2.

Dhar, O.N., Mandal, B.N. and Kulkarni, A.K. (1985): Comparison of areal raindepths of most severe rainstorms of Gangetic West Bengal and North Bengal, Proceedings of National Seminar-cum-workshop on Atmospheric Sciences and Engineering, Jadavpur University.

Gumbel, E.B. (1954): *Applied Mathematics* Series, No.33, National Bureau of Standards, Washington.

Hershfield, D.M. (1961): Estimating the probable maximum precipitation, *Journal of Hydraulics Division*, American Society of Civil Engineers, Vol. 87 (Hy.5).

Hershfield, D.M. (1965): Methods for estimating probable maximum precipitation, *Journal of American Water Works Association*, Vol.57, No.8.

India, 1996: A Reference Annual, Publication Division, Ministry of Information and Broadcasting, Government of India

Mandal, B.N., Rakhecha, P.R. and Ramana Murthy, K.V. (1985): Time distribution of rainstorms over the Indian region, *Hydrology Review of HILTECH*, NIH, Roorkee, Vol.11.

Probable Maximum Precipitation Atlas (1989): Indian Institute of Tropical Meteorology (IITM), Pune.

Srivastava, S.N. and Mandal, G.S. (1995): High rainfall during 24 hours in India, *Mausam*, Vol.40, No.1.

Wiesner, C.J. (1970): *Hydrometeorology*, Chapman and Hall, London.

World Meteorological Organization (1973): *Manual for Estimation of Probable Maximum Precipitation*, WMO, No. 332.

23

Land Degradation, Environmental Impacts and Fluvial Geomorphology: Exploration of the Linkage in the Tropics and Subtropics

Avijit Gupta

Geomorphic processes, contrary to the old textbooks, operate in extremely dynamic fashion over a large part of the tropics and subtropics. This is expected to be the case near plate boundaries or along the common tracks of large tropical storms, but destruction of natural vegetation and intensive land utilisation has increased the rate of erosion and sediment transfer to a very high level across the tropical lands. Even in places where the natural rate of erosion is low, as under the rain forests, progressive degradation of land cover currently contributes an enormous amount of sediment to the local channels. The majority of the case studies that demonstrate this rise in erosion and sediment transfer necessarily cover small areas. Publications dealing with the erosion and sedimentation at the global scale appear at intervals and generate great interest (Holeman, 1968; Milliman and Meade, 1983; Walling and Webb, 1987; Milliman and Syvitski, 1996). However, there are not many of them, as it is difficult to generalise at this scale. It is worthwhile to examine the relationship between land degradation and accelerated erosion at a large scale. This discussion, therefore,

(1) evaluates the application of fluvial geomorphology in determining the environmental impact of land degradation, (2) illustrates the use of remote sensing in studying the effect of land alteration over a large area, and (3) reviews two case studies where geomorphologists are in a position to provide valuable information of the linkage between land degradation and environmental impact for very large rivers.

The two case studies are on the Yangtze and the Mekong, and illustrate the need for involving earth scientists vigorously in studies of this type.

Land Degradation, Fluvial Geomorphology and Environmental Management

Land degradation is a term used to cover a broad range of negative environmental impacts, some of which are geomorphic. The effects could be multifaceted as in the case of the unfortunate attempt to drain the wetlands of the Barito river in Kalimantan for extending paddy cultivation and settling migrants. The ill-advised project, along with other aspects of environmental degradation, led to the draining of peat swamps leading to changing the acidity of the water in the rivers with a reported disastrous consequence on the riverine flora and fauna.

Environmental management requires both identifying problems and devising techniques for their solution. Broadly speaking, two different types of situations may arise. In the first case, the environment has already been degraded by an anthropogenic act such as extraction of timber from a rain forest or establishing a settlement on a steep hillslope. Management here involves application of our knowledge towards ameliorating an existing difficult situation. In the second type, a proposed project is assessed for possible environmental impact before it is planned in detail, i.e., at the pre-feasibility stage. Under such circumstances, opportunities exist for redesigning the project so that the probable environmental impacts are lessened considerably, but the beneficial effects remain.

Unfortunately, in spite of some excellent work, geomorphologists still play a limited role in making the world a better place to live in. This is because of (1) the limited contributions so far from the practitioners, and (2) the slow recognition of the ability of geomorphologists to contribute beyond the academic world. A large number of studies that exist, deal with accelerated erosion and sedimentation, consequences of constructing large dams, and urbanisation-related problems.

A number of case studies exist on the geomorphic effects of deforestation (Ruangpanit, 1985; Lai, 1993; Douglas, 1996). Removal of forest cover and disturbance due to felling leads to increased slope failure, slopewash, and gully formation. Intense tropical rainstorms speedily convert the logging roads to channels for transferring water and sediment to local streams. Annual sediment yield from undisturbed forests usually is 10^2 ton/km^2 when the forest is cleared. Rapp et al (1972) showed an increase from almost nil to 10800-17000 tkm^{-2}yr^{-1} following forest clearance in Tanzania. Anthropogenic impacts on small watersheds in the central Himalayas have been reported to increase the denudation rate by a factor of 5 to 10 (Rawat and Rawat, 1994). Accelerated erosion and sediment transfer diminishes after several years as the ground acquires a new vegetative cover, but if logging continues in the area, the supply of sediment to the channels remains high. This sediment partially fills in first and second order channels and makes them flood-prone. If deforestation continues, channels and floodplains of major streams are also affected. Given time, the sediment ultimately reaches the coast to interfere with normal beach building processes, creeks, and mangrove and coral communities. This, however, takes time and also continuation of the deforestation, unless the deforested slopes are close to the sea.

Accelerated sediment production is caused also by agricultural expansion, especially on steep hillslopes. Sediment yield can be high as estimated annual figure of 4000 tkm^{-2} from the Cimanuk basin in western Java shows (Aitken, 1981). This leads in the same way to channel modification, reservoir silting and building of coastal plumes. The annual sediment yield from the Himalayas has been estimated as about 10^3 tkm^{-2} (Ives and Messerli, 1989). The effect of this sediment appears to be local, and most of it is probably stored on the foot-slopes (Bruijnzeel with Bremmer, 1989), and gradually transported down rivers. Obviously, here is role for geomorphologists for measuring the degradation and planning their avoidance.

The construction of a very large number of dams in the world in the early second half of the twentieth century has led to environmental degradation. According to the International Commission on Large Dams (ICOLD), 35000 large dams (each more than 15 m high) exist worldwide, not counting China. The trend has also been towards building big dams. Dams have a number of beneficial impacts but the list of potential environmental degradation is also long. Geomorphologists need to study the effect of sediment accumulation in reservoirs, the changed downstream flow pattern, the narrowing and deepening of the channel immediately downstream of the dam, loss of sediment supply to downstream coastal areas leading to coastal erosion, and problems of land subsidence and slope failures. The effect can be both local and distant. The front of the Nile delta is undergoing severe erosion following the closure of the Aswan dam. Similar problems have been reported at the mouth of the Volta river and have been associated with the large Akosombo dam of Ghana. Given the number of dams across the rivers of the world, the cumulative starving of sediment of the coast may turn out to be crucial, especially if and when the climate-driven sea level rise occurs. This is an opportunity for geomorphologists to be useful to the world.

About half of the world's population currently lives in urban settlements, the number which is expected to rise to about 60% by 2025 (United Nations, 1995). Most of this will happen in the less developed countries, almost all of which are located in the tropics and subtropics. Cities of all sizes are increasing in number, especially cities that carry between 1 and 5 million people. The other trend is the rise in the number of megacities, most of which, unlike the past pattern, are now in less developed countries. Urbanisation by itself may lead to flooding and accelerated sediment production. The physical problems are compounded by the fact that a number of urban settlements are located in hazardous zones such as near active plate margins or common tracks of tropical storms. The problem of subsidence also has started to trouble the cities on coastal plains and deltas that depend on extraction of water from the subsurface. Management of the urban environment due to land degradation has already become difficult (Gupta and Ahmad, 1999).

The Large-Area Image

Most of the work on the application of fluvial geomorphology on measuring land degradation, environmental impact assessment, and designing ameliorating measures has been necessarily of small scale. A significant part of the data come from experimental watersheds (≥ 10 m^2), first or second order basins, or hillslopes of limited length. Data at this scale require care in extrapolating the numbers over large areas. Some measurements come from gauging stations on rivers such as the Cimanuk which collect sediment concentration amounts but such stations do not form a dense network and generalisation over large areas remains problematic. The data from these stations, however, does provide information that represents the entire basin lying upstream of the station. The worldwide inventory of sediment released to the coastal waters, referred to earlier, to a great extent is derived from this kind of information.

The very large collection of satellite scenes archived at various ground stations across the world, forms a scarcely used source for identifying land degradation and sequential problems. For all we know, such scenes may run into millions, a large part of them with excellent resolution, and scenes of the same area can be studied in a time sequence. These can be used over a wide range of scales. Given that recently satellite images have become available at a resolution which is almost a metre, these can be used also for detailed analysis, provided one has the funding to acquire them.

Several years ago, my colleagues at the National University of Singapore and I mapped coastal sediment plumes for parts of South and Southeast Asia (Gupta and Krishnan, 1994; Gupta, 1996). The mapping was done with the AVHRR imagery (resolution 1.1 km) and Landsat MSS and TM (resolution 30 m). We have subsequently extended the mapping with SPOT images (resolution 20 km). We found the following:

1. It is possible to map sediment plumes in coastal waters even from AVHRR imagery, which is very useful for mapping extremely large areas.

2. Mapping at better resolution provides information regarding the direction of plume movement.

3. It is possible to monitor the nature of the plume over seasons and over years.

4. A geomorphologist can identify plumes that are incongruous with the physical environment.

5. The incongruous plumes provide information on where land degradation is reaching serious proportions.

6. Types of coastal environment (mangrove, coral reef, beach) vulnerable to the arrival of excessive sediment could be identified.

7. The regional pattern, matched with sediment data from experimental stations can be used for environmental management at various levels.

8. The aerial extent of the coastal plumes provides a semi-quantitative measure of the sediment being transported but the actual amount has to be measured in the field.

The Centre for Remote Imaging, Sensing and Processing (CRISP), Singapore has been studying the middle Mekong basin for the last three years using SPOT images in order to establish how much descriptive and interpretative information can be derived from such scenes. Recently, there was an opportunity to do some ground checking which turned out to be extremely encouraging. This project has been briefly described earlier (Gupta, 1998; Chen et al. 2000) and a detailed account is currently under preparation. A discussion of the Mekong is included in a later section of this paper. It has been possible to identify the sources of accelerated sedimentation and the excessive bar building in this large river.

Although the examples provided here exclusively refer to accelerated erosion and sediment production, I would like to submit that:

1. It is possible to examine land degradation and other kinds of environmental problems at a regional scale and repeatedly over time from satellite imagery.
2. This work can be done visually by a good geomorphologist familiar with the area.
3. Numerical estimations are possible when image interpretation is associated with some ground truthing and measurements on ground are available.
4. This database of high potential is very little utilised, which is surprising given the speed and accuracy with which a large area can be mapped.
5. The availability of images at resolution of about a metre is an excellent opportunity for detailed studies of specific objects.

The linkage between fluvial geomorphology and environmental management in two large drainage basins are shown in the following sections. The example from the Yangtze is used to illustrate the need to look at environmental issues with a large-scale perspective. The second case study on the Mekong also includes a demonstration on application of satellite images.

The Yangtze (Changjiang) River

Meade (1996), in his list of the very large rivers of the world, identifies the Yangtze as the 5th largest in terms of mean annual water discharge (900×10^9 m^3 of water). In terms of average annual sediment discharge it is the 4th largest (480×10^6 t). Four large rivers rise on the Tibetan Plateau and in close proximity: the Yangtze, the Song Hong (also known as the Red), the Mekong, and the Salween. Rising at elevations near 5000 m, all the rivers flow through structure-guided gorges to the south and the southeast. The Yangtze, which is known as the Jinsha for most of its upper reach, flows south to Shigu where it abruptly turns north by northeast. Several more sharp turns take the Jinsha to Chongqing where the Jialing joins it from the north. The Yangtze, which in

China is known as the Chenjiang, then flows northeast through the Sichuan Basin in a structure-guided course, and then east through three huge rock gorges to Yichang where the river begins to meander eastward across a wide alluvial basin imperfectly filled with sediment. This is the Yangtze flood hazard zone and, in the earlier days before the construction of the flood dykes, part of the water was retained in a number of lakes and depressions such as the Dongting and the Poyang lakes. The huge river turns east and northeast, past Wuhan and Nanjing to build a large delta in the East China Sea near Shanghai. The river is more than 6300 km long and has a total drop of about 5000 m. Some snowmelt feeds the Yangtze in its upper course but it really is a monsoon river that rises in summer and floods following episodic heavy rainfall caused by tropical storms. Intense land cultivation has silted up its natural flood-storing arrangements in the wetlands. The river also receives a large sediment load from the steep slopes of its upper course, both due to natural and anthropogenic causes. The sediment can no longer be spread over the floodplain, and is deposited on the bed of the dyke-bound channel or taken out to the delta.

I am not going to discuss the effect of the Three Gorges dam on the Yangtze in detail. That has been done in many places and a vast literature is building up on the expected benefits and disasters that may arise when the project is completed in 2009. Most of the criticism of the project involves silting in the reservoirs, problems of flood retention, and the effects of large-scale resettlements. It will inundate 13 cities, 140 towns and 1352 villages, causing a resettlement of 1.25 million people. It has been hypothesised that resettling that large number from the lower slopes to the higher slopes above the level of the reservoir will contribute to even greater amount of sediment than the river now receives.

Although the discussion on environmental degradation has mostly focused on the Three Gorges area, problems arising out of lack of sediment downstream of the dam have been identified. Rivers below a large dam usually scour its bed creating problems of channel stability and functioning of river-related structures such as bridge piers or canal intakes. The decreasing amount of water and sediment arriving at the delta face may create problems for Shanghai, allowing the saltwater wedge to move upstream. The problem may be further accentuated especially in the dry water season, if the project for transferring water from the Changjiang to the water-scarce north comes into operation.

This in not the place for a detailed discussion on one of the most difficult and large-scale attempts to control a major river of the world. I only like to stress that we as earth scientists have a role to play in such complex projects and, when we do so, we should remember that the environmental issues are not necessarily limited to an area in the vicinity of the project.

The Mekong River

The Mekong is the 12th longest river (4880 km) in the world and is ranked 9th according to mean annual discharge (475×10^9 m^3). It is 10th in terms of average annual suspended sediment discharge (160×10^6 t). It flows in a pan-shaped basin that drains into the South China Sea. Rising in the highlands of eastern Tibet, the Mekong flows through six countries: China, Myanmar, Lao PDR, Thailand, Cambodia and Vietnam. The river flows through a structure-guided rocky course interrupted with limited reaches on alluvium until inside Cambodia where it has built a large alluvial embayment in the structural depression of Tonle Sap, and further south, a large delta. It is also a seasonal river, which may rise in floods, especially in Cambodia and Vietnam as last year's events demonstrated on a massive scale. The combination of seasonal pattern of discharge with structural lineation-guided tributaries and steep hillslopes of the upper and middle basin has the potential for considerable environmental degradation, especially following forest cover removal and anthropogenic slope and channel modification. For decades, with the cooperation of the member riverine states, development of the Mekong basin has been planned in a structured fashion by an international organisation, the Mekong River Commission, and its predecessor, the Committee for Coordination of Investigations of the Lower Mekong River Basin. Land in the basin is either under some form of agriculture or under forest. A considerable disparity does occur among the different parts of the basin regarding the intensity of landuse and population density, which varies between 8 persons per km^2 in the highlands and 570 in the fertile delta.

At the CRISP, Singapore we investigated 13969 km^2 of the Mekong basin in Lao PDR which was covered by four SPOT scenes. We intended to demonstrate that using visual interpretation, we could:

1. prepare a geomorphological account of the area concerned,
2. identify the zones which are vulnerable to environmental degradation, and
3. examine the area for present and future environmental impact of the landuse changes and the proposed development projects, as mentioned in the various publications of the Mekong River Commission.

We mapped the study area in terms of environmental vulnerability and the Mekong channel in terms of sediment transfer and deposition. We also identified areas, pixel by pixel, which seem to contribute sediment to the channel because of their location on steep slopes that have been cleared of vegetation. Some of that is due to clear felling, but the common practice of shifting cultivation on steep slopes, growing hill paddy and vegetables seems to cover a very large area and is possibly the important ongoing source of sediment supply to the channel. It was also possible to trace the passage of the sediment by episodic transfer and storage via the tributaries and direct slope failure to the Mekong channel. The seasonal pattern of storage and transfer operated in the main channel also, but in the Mekong the locations of rock protrusions act as core areas of bar accumulation in fresh sand. The Mekong has some inset bars, but its mid-channel bars are wrapped round a rocky core. The information has allowed us to suggest steps for environmental management in this part of the basin and also for the problems that will be created by constructing any control across the river. Any reservoir on the Mekong will have an astonishingly short life span. Having located a number of images between 1992 and 2000, we are now tracing origin and accumulation of sediment over time and also its seasonal variation in a particular year. We were encouraged by the fact that our visual interpretation of the satellite images was borne out in the field.

This is an example of the profitable combination of remote sensing and fluvial geomorphology in understanding a river system and also a quick and large-scale evaluation of environmental vulnerability and potential impact of proposed development projects. Given the large areas that need to be studied within a finite period of time, remote sensing, even with simple visual interpretation, is an extremely efficient tool. We took only several weeks to map almost 14000 km².

Conclusion

Our knowledge of fluvial geomorphology has now reached a stage where it can be used for environmental impact analysis and management. Usually, we study the relationship between land degradation or similar misadventure and environmental deterioration over a small area. Given the scale of the environmental modification and the size of some of the projects in the tropics and subtropics, we should also start looking at the large-scale scenarios. The global archive of innumerable satellite images collected over years is a much-neglected tool, even when interpreted visually. It is, however, an excellent practice to check the conclusions reached later in the field.

References

Aitken, A.P. (1981) Aspects of erosion and sediment transport in Java, Indonesia In: T. Tingsanchali and H. Eggers (Eds.) *Proceedings, Southeast Asian Regional Symposium on Problems of Soil Erosion and Sedimentation,* Asian Institute of Technology, Bangkok, 81-91.

Chen, P., Lim, H., Huang, X., Gupta, A. and Liew, S.C. (2000) Environmental study of the middle Mekong basin using multi-temporal SPOT imagery, paper presented at the IGARSS conference, Hawaii, 2000.

Douglas, I. (1996) The impact of landuse changes, especially logging, shifting cultivation, mining and urbanization on sediment yields in humid tropical Southeast Asia: A review with special reference to Borneo In: D.E. Walling and B.W. Webb (Eds.) *Erosion and Sediment Yield: Global and Regional Perspectives*, International Association of Hydrological Sciences Publication 236, Wallingford, 463-471.

Gupta, A. (1996) Erosion and sediment yield in Southeast Asia: A regional perspective In: D.E. Walling and B.W. Webb (Eds.) *Erosion and Sediment Yield: Global and Regional Perspectives*, International Association of Hydrological Sciences Publication 236, Wallingford, 215-222.

Gupta, A. (1998) Rapid erosion risk evaluation in the middle Mekong basin by satellite imagery, *Proceedings, Euro-Asia Space Week*, Singapore, ESA SP-430, 241-245.

Gupta, A. and Ahmad, R. (1999) Geomorphology and the urban tropics: Building an interface between research and usage, *Geomorphology*, 31, 133-149.

Gupta, A. and Krishnan, P. (1994) Spatial distribution of sediment discharge to the coastal waters of South and Southeast Asia In: L.J. Olive, R.J. Loughran, J.A. Kesby (Eds.) *Variability in Stream Erosion and Sediment Transport*, International Association of Hydrological Sciences Publication 224, Wallingford, 457-463.

Holeman, J.N. (1968) The sediment yield of major rivers of the world, *Water Resources Research*, 4, 737-747.

Lai, F.S. (1993) Sediment yield from logged, steep upland catchments In: J.S. Gladwell (Ed.) Hydrology of Warm Humid Regions, International Association of Hydrological Sciences Publication 216, Wallingford, 219-229.

Meade, R.H. (1996) River-sediment inputs of major deltas. In: J.D. Milliman and B.U. Haq (Eds.) *Sea-Level Rise and Coastal Subsidence: Causes, Consequences and Strategies*, Kluwer Academic Publishers, Dordrecht, 63-85.

Milliman, J.D. and Meade, R.H. (1983) World-wide delivery of river sediment to the oceans, *Journal of Geology*, 91, 1-21.

Milliman, J.D. and Syvitsli, J.P.M. (1996) Geomorphic/tectonic control of sediment discharge to the ocean: The importance of small mountainous rivers, *Journal of Geology*, 100, 525-544.

24

River Bank Erosion and Land Degradation: A Study on River Ganga

Sutapa Mukhopadhyay

Land is the primary resource of the world. Land degradation, expressed as soil erosion, is a well-known phenomenon and appropriate measures have been taken against it over many centuries. The most generally occurring kinds of soil erosion are water erosion and wind erosion, these being the most active agents on the surface of the earth. Stream-bank erosion is a special kind of erosion which is responsible for huge land loss of the world. Land degradation is a crucial issue in the riverine areas due to undercutting at the base of the bank.

The bank erosion is frequent just downstream from the axis of the bend, with both bank caving and point bar deposition. Concentrated slightly downstream from the bend axis, river curves tend to move down the valley. The raising of meandering belt may give rise to 'avulsion', vast changes in the line of meander trains (Petts, 1985). Channels continue to receive a permanent input by seepage through the permeable embankments and overspill during floods. Bank erosion is prominent because the river seeks to return to her natural course. Thus, after channelisation the course has extended, as seen in the bank of Mississippi river where the rate of erosion is 900,000 m³ every year (Petts, 1996).

For thousands of years rivers have been altered by many forms of human activity. Channelisation is an ancient practise to regulate flood water. The direct impact of this physical regulation of the river is related to the raptured connectivity between the functional units of the river system and it has markedly altered the flows, sediment loads, and water quality characteristics of many rivers. Rivers can also dissipate their energy by channel enlargement through lateral erosion of the banks, in particular if the bed is armoured. Lateral erosion results in increased sinuosity, lengthening of the stream and the reduction in gradient and stream velocity (Leopold, 1964).

Despite the natural processes some fundamental changes have been brought about by the human interferences on the natural flow of the river. By far, the most serious environmental crisis facing poor countries like India is the loss of land on which most of the people depend directly for their basic sustenance. Thus, in India, where 72% of the people still depend on agriculture for their livelihood, the basic developmental planning must be the conservation of land resource from any type of loss.

India is par excellence a land of rivers and river basins. The total discharge per year of all the river basins, as estimated by Central Water and Power Commission, based on Khoslas formula, is 208,600 million m^3 and according to K.L. Rao it is 475,100 million m^3 (Rao, 1979). In a country like India where agriculture is the mainstay for an overwhelming majority of population, after independence attention was paid towards developing the surface water sources from the existing drainage system. So, to protect the agricultural land from floods as well as for supplying the irrigation water, large number of dams, barrages, embankments have been constructed along the rivers in various parts of the country as a part of the river basin management.

The Ganga is the main river of India and the last course of this river is in West Bengal state, eastern India. After following 76 km from west to east over West Bengal, she enters Bangladesh after crossing the international boundary. Farakka barrage has been constructed in this sector to regulate the flow. But, the devastating soil erosion and havoc flood of the river Ganga have become an annual feature of Malda and Murshidabad districts of West Bengal. More or less five community development blocks in Kaliachak, Manikchak and Ratuna police station in the district of Malda have been suffering a lot due to erosion of soil and flood caused by the river Ganga. It is estimated that about 750 sq km of land including cultivated land, dwelling houses, mango orchards and gardens have been lost in a couple of years to the river Ganga due to fatal and devastating bank erosion. So, in this paper emphasis has been given on this specific land degradation problem, its probable cause and nature along the 78 km long stretch of lower Ganga.

Huge volume of water (about 46 thousand crore cusec) from the catchment area of about 10.50 lakh km^2 passes through this stretch of the Ganga river. The significant factor is that 80% of the total volume flows during three rainy months, so spilling over bank and flood is common in this tract. Locally this lowlying area subjected to inundation with rise of the river level is known as Tal and Diarha. The dark clay soil is highly fertile and suitable for mango orchards, which has a good international market. The mulberry plantation is also practised here. The formation of these terrain units is the result of centuries of fluvial action of the Ganga, and the irony is that these lands are now eroded away also by the action of the Ganga herself.

Land Degradation Along the Bank of River Ganga

Nature

Though there is no concrete report of land erosion along this stretch before 1970, but following Keshkar report it can be stated that from 1931 to 1971 the total amount of land erosion of this part was 14,335 hectares. After the construction of the barrage the rate of land erosion accentuated largely and within a single year (1997-98) it was 1400 hectares of land. In 1972, some measures were taken by the government against land erosion but the results were nil.

Amount of Land Erosion in Malda District

Year	Land erosion (hectare)
1931-71	14,335 (average 716 hect/year)
1972	729
1973	939
1974-78	868
1979-95	22,160 (average 1385 hect/year)
1996	1310
1997-98	1500
1998-99	1720

It is estimated that about 10 million m³ soil is mixing in the river Ganga due to severe erosion every year (Rudra, 2000). It is very painful to say that there is no existence of 60 primary schools, 14 high schools, BDO office, Anchal office, temples, mosques, etc., which have all been snatched by the Ganga during the last 30 years. According to the last census report, 47 villages out of 89 in Manikchak block, 17 villages out of 66 in Kaliachak-II and one kilometre land in Tofi and Birnagar area have been lost to the river Ganga in a couple of years. In the year 1999, fresh erosion has started from Manikchak to Kaliachak-II, the sixth retired embankment at Khaskol and seventh retired embankment at Mohanpur were breached and about 1530 hectares of land are eroded away (Report, 1999).

Cause

(i) In response to the analysis of the land erosion, the centuries-old changing course of the river Ganga can be discussed. Satellite pictures show that the Ganga is in braided form between Maharajpur and Rajmahal. The construction of the boulder embankment surrounding the Bhutni char and the Fulahar embankment restrict the natural deposition of sediments in the island; as a result the silt is deposited along the left channel (Mora Kosi) and now it has become almost silted up. So, the main stream is now flowing along Rajmahal. Due to the presence of hard granitic rocks of Rajmahal, the stream gradient as well as the velocity of the flow increases and it strikes the opposite bank at a high speed directly at 45°. This is the main cause of the rapid east-bank erosion of the Ganga and the gradual shifting of the river towards Malda, instead of flowing southwards to Farakka which was the original flow (Bigyan Mancha, 1999).

(ii) To protect the land from erosion, in 1972 marginal embankments were constructed but from 1972 to 1991 these embankments were breached nine times at Simultola. In 1994 at Aswinitola, 25 kms upstream of Farakka barrage, fourth retired embankment was constructed. In 1995, one km of it was breached and in 1996 nearly 3 kms of it was breached and the whole flood water entered Malda town through this opening. It is also observed that from 1992-96 there his been a 2 kms shift at Malda town. According to the 1999 report, the river is shifting 600-750 metres annually along this stretch. In 1997, a retired embankment was constructed at Gopalpur, 800 m far from the existing river bank but in 1998 this gap reduced to only 40 metres. At the same time, the gap between the Pagla, Bhagirathi and the Ganga is reducing at an alarming rate and now (September, 2000) it is only 170 m. River geomorphologists predict that in near future the Ganga will flow through Pagla river and this will lead to a state when Farakka barrage will remain as a bridge only. The history depicts that upon 13th century Kalindi, Bhagirathi and Pagla were the old courses of the Ganga in different ages, which now remain as abundant channels along the left bank of the present Ganga. It is also observed that by eroding the left bank it has a tendency to flow through these old courses.

(iii) The construction of Farakka barrage is supposed to be another important cause of this land erosion. Almost 54 sluice gates by the side of the Malda district have been rendered unworkable, so about 18-27 lakh cusec water cannot flow through this barrage during rainy season. As a result, the flow of the river Ganga has been obstructed in these sluice gates and a long strip of land, 910 kms long and 213 kms, wide has emerged in the upstream of the barrage. Due to this obstruction severe erosion takes place just upstream of it (Rudra, 1998).

Moreover, due to faulty construction, the spurs for the control of the river flow have been fully damaged and washed away. As a result there is an increase in land erosion day by day and deposition of spur materials behind the barrage. In 1999, 24 no. spur worth 13 crore rupees was completely destroyed before completion and about 1720 hectares of land was also eroded away and 42 thousand tons boulders mixed up in the river bed.

The river is a living reality that needs space. Deltaic rivers shift constantly, veering from an extreme right bank point to another on the left bank, maybe over centuries, perhaps over decades. These extremities are called the 'reach points'; on the river banks these are the lines of lineament (Warner, 1985). At present the reach point of the Ganga at Malda is several kilometres from the present bank line and probably stretch to Englishbazar area, the heart of the town. And the line of lineament is probably at the point where the proposed spurs no. 28 and 29 are to be built. These two spurs fall in the permanently eroding zone. But, ignoring this river dynamics the local politicians demand the construction of spurs no. 28 and 29 immediately, then who will convince them and will show them the future? They want quick solution for quick result.

Proposals for Mitigation

Inundation is a natural phenomenon; what we forget is that the entire area in which we have developed our homes and cultivation fields, the Bengal delta, is the creation of annual inundation over millions of years. It is our greed that forced us to settle right on the banks, reclaiming land permanently, against the wishes of nature. Now we have forced the river to rebel and bank erosion is the expression of such rebellion.

The irrigation and waterways departmental report on "Erosion Problem of Ganga Upstream of Farakka Barrage in Malda District" (7(4) Part A, 1998), stated that the river Ganga should be allowed to move to the line of lineament (its natural reach point, up to which it is bound to erode) so as to allow the flow to spread up to the Malda-Manikchak road, and also to allow erosion to take a natural course for the year 2002 AD. This a serious point and perhaps this is the best solution.

Some others measures that could be taken are:

1. Dredging two channels, one to the right of Bhutni Diara island and other along the middle of Rajmahal channel.
2. Buttressing the existing main channel on the extreme right.
3. Dredging the Koshi channel.
4. The present course of river Foolhar should be changed in as much as the flow of its water should be linked and/or connected with the flow of the river Ganga which was the original course of the river Foolhar. It is to be noted that in the recent times a long strip of land has emerged in the Ganga near about the mouth of the Foolhar, and the water of the Foolhar cannot flow to the Ganga due to the emergence of this shoal.
5. Long strip of land in the middle part of the channel, about 10 kms in length and 2-3 kms in width, formed at the upstream of the barrage on the left side of the Ganga should be removed.
6. Erosion of soil mostly takes place 20 to 26 km upstream of the barrage so the spurs should be constructed 28-29 km from the barrage as suggested by Pritam Singh Committee and CWPRS, Pune.
7. There is a need for better study of the new morphological changes of the river.

Conclusion

Land degradation has become a problem of existence for the villagers of this area. They have organised Gana Namaz (Muslims' prayer) and prayed 'Oh Ganga Ma! Please spare us, do not eat up more of our land'. Billion-dollar solutions have not solved the problem, nor have the government and international organisations or the involvement of the people's participation in local issues.

In the end, it may be said that it is wise to accept the greatness of huge magnitude of nature and to understand the dynamism of it, scientific study is essential. Planning should be outlined depending on this changing nature of the river otherwise people will have to spend all their resources only to defend themselves from natural hazards.

References

Leopold, L.B., Wolmen, M.G., Miller J.H. (1965) *Fluvial Processes in Geomorphology*. Eurasia Publishing House, pp. 305-307.

Petts, G. and Foster, I. (1985) *Rivers and Landscape*. Edward Arnold, pp. 140-170.

Petts G.E. (1996) *Fluvial Hydrosystems*. Chapman and Hall, pp. 75-95.

Rao, K.C. (1979) *India's Water Wealth*. Orient Longman, pp. 40-45.

Report of Ganga Bhangan Pratirodh Committee (1999), Malda district, West Bengal.

Rudra, K. (1998) Navigability of Calcutta Port and Ganga Accord of 1996. Parivesh O Unnayan. Akhil Bharat Bhuvidya O Parivesh Samiti, Bolpu (An article in Bengali).

Rudra, K. (2000) Changing Course of the Ganga. *Ananda Bazar Patrika*. 20th September, 2000.

Warner, R.B. (1985) Spatial Adjustments to Temporal Variations in Flood Regime in Some Australian Rivers. *River Channels* by K. Richards. Basil Blackwell, pp. 19-27.

West Bengal Bigyan Mancha (1999) Bulletin of Round-table Conference. 15-16th May, 1999.

25

Flood and Land Degradation: A Study in Lower Ajoy River Basin

Mallanath Mukherjee and Malay Mukhopadhyay

In India, flood hazard is a spatially extensive phenomenon and all the states are affected in varying degrees. The most severely affected are Andhra Pradesh, Orrisa, Uttar Pradesh, Bihar, Punjab and West Bengal, in terms of total value of damages. Floods in India cause damage worth of Rs. 1000 crores annually and the major damage is to crops, accounting for 73% of the total, the rest is to houses, livestock and public utilities. It may be pointed out that there is constant increase in the frequency, intensity, spatial dimensions, magnitude and damage, and above all nature of damages, by floods in the present time. Besides this, another major effect is land degradation in terms of erosion of fertility and productivity. During rainy season thousands hectares of land is eroded due to rapid flow of water along the river Bhagirathi and river Padma. This has become a threat to physical as well as socio-economic environment of the nation. The flood situation and its adverse effects in some parts of eastern India, particularly in Lower Ganga plain, regarding bank erosion and land loss have attracted the interest of national leaders, planners as well as geographers.

As land is an important component of physical environment, land degradation therefore causes environmental degradation in any ecosystem. Land degradation means overall lowering of land qualities because of adverse changes brought about by human activities and some physical events. Land degradation reduces soil fertility, agricultural productivity and gross agricultural output, and results in decrease of biological diversity of the natural ecosystem. In the present study, flood has been shown as a causative factor for the occurrences of land degradation in the lower Ajoy river basin of West Bengal. In this area, large scale sand splay at several points occurs in the river astride agricultural lands during devastating flood period which makes the land unproductive and expedites desertification in post-flood period.

Flood of Lower Ajoy River Basin

The Ajoy is one of the most important right-bank tributaries of the Bhagirathi that emerges at Bihar from Chakai hills at an altitude of 330 m, draining its flow to Bhagirathi at Katwa in Burdwan district of West Bengal. It has a length of about 260 km and an area of about 6221 km^2 of which about 2675 km^2 is in West Bengal. The lower part of the basin has been determined from Pandebeswar that lies below 80 m contour. Some important settlements like Pandabeswar, Illambazar, Dubrajpur, Bolpur, Guskara, Natunhat, Ketugram and Katwa have taken shape within its lower part over decades.

Flood has become a consistent and recurring features of this basin area. The river trickles through the sand from the month of November to mid-June and it suddenly becomes flooded due to outburst of monsoonal depression during August and September. The mean annual discharge of Ajoy has been recorded to about 65 cumecs at Natunhat. But there is much fluctuation, ranging from one cumec in April to about 250 cumecs in rainy season. The peak flow discharge figure during floods is abnormally high, which is even over 8000 cumecs.

In the present century high floods occurred in this basin in 1916, 1942, 1946, 1956, 1959, 1970, 1971, 1978, 1984, 1995, 1999,

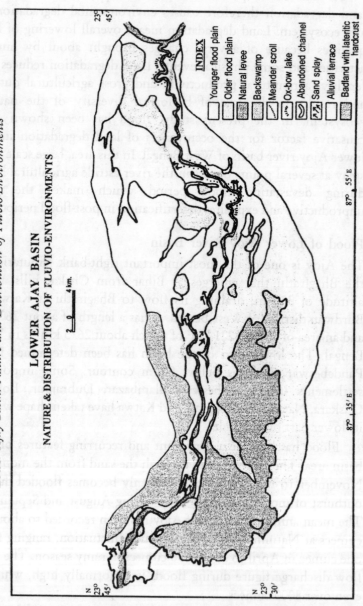

Figure 1

Lower Ajay Basin: Nature and Distribution of Fluvio-Environments

Figure 2
Lower Ajoy Basin

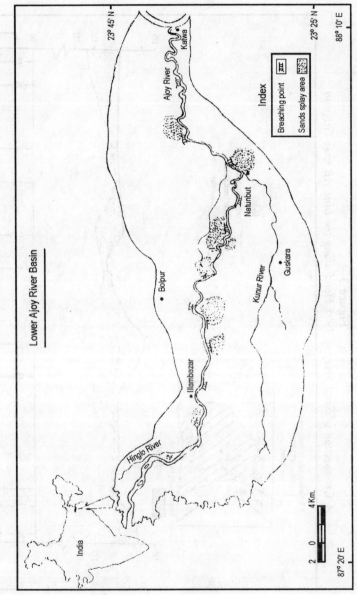

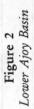

Lower Ajoy River Basin

23° 45′ N

23° 25′ N

88° 10′ E

Ajoy River

Katwa

Bolpur

Illambazar

Hinglo River

Natunbut

Kunur River

Guskara

India

Index

Breaching point

Sands splay area

4 Km.

2   0

87° 20′ E

Figure 3
A Cross-section of Sand Splay of 1995 Flood along the Line AB near Gitgram

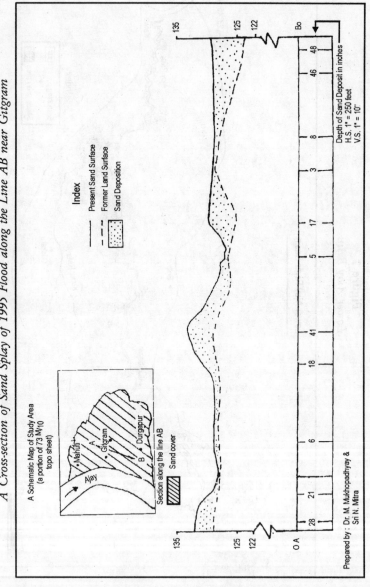

A Schematic Map of Study Area
(a portion of 73 M/10 topo sheet)

Index

— Present Sand Surface
--- Former Land Surface
▦ Sand Deposition

Section along the line AB

▨ Sand cover

Depth of Sand Deposit in inches
H.S. 1" = 250 feet
V.S. 1" = 10"

Prepared by : Dr. M. Mukhopadhyay &
Sri N. Mitra

Figure 4
A Cross-section of Sand Splay of 1999 Flood along the Line AB near Gitgram

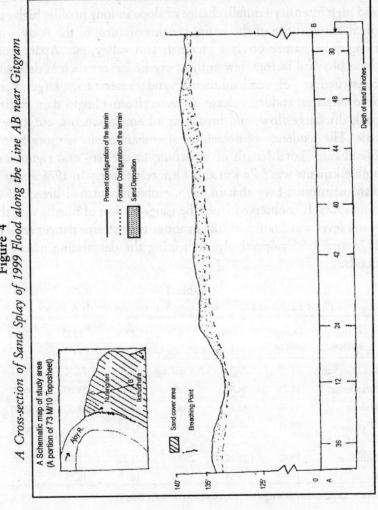

Present configuration of the terrain
Former Configuration of the terrain
Sand Deposition

A Schematic map of study area
(A portion of 73 M/10 Toposheet)

Sand cover area
Breaching Point

Nutangram
B
Iskrudhara
Ajoy R.

Depth of sand in inches

A   36   12   24   42   40   44   48   30   B

140
135
125
0

and 2000. The floods of 1956, 1978, 1995, 1999, and 2000 have been considered as notorious in terms of losses. It is observed that a number of physical and anthropogenic factors are totally responsible for the occurrences of floods in this area. Among the physical factors which cause river floods, important are: prolonged and high intensity rainfall; change of slope in long profile; high rate of siltation; shape of the basin; synchronisation of the Ajoy with Bhagirathi; nature of river channels and valley, etc. Apart from these physical factors, few anthropogenic factors such as defective construction of embankments and reservoirs, large scale deforestation, sudden release of water from Hinglo dam during peak discharge flow, and breaching of embankments, etc. play a role. The tendency of breaching of embankments is more below Illambazar. Total length of breaching in the left- and right-bank embankments was 5.78 km and 3 km respectively in 1978 and it is also mentioned here that in 1995, embankment had breached in nine spots. It is observed from the gauge station of Bhedia that the water level was 2.6 m and 2.83 m above the extreme danger level in 1978 and 1995 respectively, indicating the devastating nature of floods.

Table 1
Flood Levels at Different Gauge Stations on the Ajoy River

Gauge station	Danger level (m)	1956	1971	1978	1995	1999
Pandabeswar	74.50	77.33	No spilling	N.A.	76.40	76.48
Satkahonia	51.75	54.41	N.A.	55.44	55.00	55.20
Maliara	41.70	45.37	4.47	44.97	45.00	45.50
Gheropara (Bhedia)	39.41	40.89	40.30	42.82	42.30	43.25
Natunhat	19.40	21.00	20.56	21.20	21.00	21.40
Katwa	14.90	15.64	15.94	16.70	16.75	17.00

Source: Department of Irrigation, Government of West Bengal.

The floods of 1978, 1995 and 1999 brought about serious damage to the physical as well as socio-economic environment, inundating nearly 350 moujas of lower basin. Some villages lying below Illambazar had been completely wiped away from the map by the devastating floods. The government report reveals that 5835 hectares of paddy were totally damaged in Bolpur, Illambazar and Nanur police stations in 1995, money value of which has been estimated to about 8.35 crores. The flood water remained stagnant for about 48 hours in parts of Bolpur, Nanur, Ketugram and Mangolkote areas. In this area, water level rose to about 15 to 20 feet from the surface where uprooted paddy, straw, of electric wires and poles and even tree branches indicated the water level during flood. The intensity of flood, rise of water level and magnitude of damage is increasing gradually.

Flood and Land Degradation

The land degradation process refers to the land erosion and lowering of land qualities by running water in different suitable places like river banks, river astride agricultural lands, forest lands, hill slopes and seas. The land degradation in the lower Ajoy river basin is solely through large scale sand splay over agricultural lands that deteriorates soil fertility and agricultural productivity and, side by side, decrease the gross cropped area. In fact, the process of land degradation is a comprehensive natural geomorphological process which operates through the chain of erosion of geomaterials, transportation of eroded materials and deposition of these materials in river beds and again during flood, river bed materials being extensively deposited over agricultural lands through breaching of embankments or passing over the banks. Therefore, in the present study land degradation is a special type of sedimentation of loose coarse grained sands with poor cohesiveness in the productive agricultural lands.

The land degradation due to sand splay is commonly found in the downstream of Illambazar. It is because of the fact that average slope of the basin area is 1 in 750 from source region to Pandabeswar, but after Pandabeswar it drops to about 1 in 2250. Due to great slope differences along the lower course, rate of siltation is very high which decreases the water accommodating capacity of the river. In addition to this, varying width of embankments on both sides, more meandering nature of the river, and high discharge of water from Hinglo dam cause breaching of embankments in the left and rights bank respectively. Since 1978, breaching points have been recorded at about twelve places. Through the breaching points large amount of riverine sands have been deposited over the agricultural lands which have damaged kharif crops in the village. A report on damage by sand splay is given below.

Table 2
Spatial Distribution of Sand Splay

Mouja/village	Police Station	Area (hectares) covered by sand splay
Bhedia	Ausgram	70.71
Brahman-Dihi	Ausgram	30.71
Mallikpur	Ausgram	28.68
Natunhat	Mongolkote	40.08
Nabagram	Mongolkote	31.77
Bira	Ketugram	26.34
Bankai	Ketugram	16.52
Narega	Ketugram	20.09
Gitgram	Bolpur	70.82
Haripur	Bolpur	35.64
Rasulpur	Bolpur	75.28
Itanda/Natungram	Bolpur	65.00

Source: Department of Panchayat and Rural Development, Government of West Bengal.

The sand splay and land degradation in and around the channel cause a series of environmental problems which have introduced ecological as well as socio-economic imbalance in the region concerned. The land degradation problems are given below:

(i) Loss of Soil Nutrients and Agriculture Production

Deposition of coarse sands in the flood plains during the period of extensive flood buries fertile alluvial soil and renders the flood plains infertile and convert it into wasteland. This large scale splay, with an average depth of about 2 to 3 feet (Fig. 2) in many places, has reduced nutrient concentration in the soil, as a result of which productivity has decreased substantially. With a view to examine soil quality, two-phase soil nutrient analysis had been made. Its results are displayed in Tables 3 and 4 below.

Table 3
Soil Status Report Before Sand Splay

	pH	Nitrogen kg/ha	Potassium kg/ha	Phosphorus kg/ha
Natungram	6.6	1160	440.8	70.5
Gitgram	7.0	1040	390.5	60.4
Bhedia	6.9	1100	410.8	65.3
Narega	7.1	1220	370.3	40.2

The study has been carried out since 1995 and soil nutrients like nitrogen, potassium, phosphorus and pH level have been examined from the above places in different periods. The results of samples collected before sand splay, indicates that pH level varies from 6.6 to 7.1 and other nutrients are favourable for food crops like paddy, wheat, etc. Generally, pH level 6.6 to 7.3 is most favourable for food crops and vegetables.

Table 4
Soil Status Report after Sand Splay

	pH	Nitrogen kg/ha	Potassium kg/ha	Phosphorus kg/ha
Natungram	7.8	510	102.6	18.25
Gitgram	7.9	480	90.8	8.60
Bhedia	7.9	470	106.7	6.30
Narega	8.1	370	110.0	5.84

The soil testing report for post sand splay period indicates that these areas are now not suitable for any food crops and vegetables, because the presence of major soil nutrients like nitrogen, potassium, phosphates is very poor. The pH level of 7.4 to 8.4 is slight to moderate alkaline and it would only favour some root crops like sugar beet, potato, and some forage crops like paspalam dilatatum, dinanath, etc. It is observed from the village level survey that before sand splay the production of paddy was about 5 quintal per bigha and at some places even less than 1 quintal. But after sand splay production has decreased to about 1 quintal per bigha, and to some extent the cultivators left the land as fallow.

(ii) Loss of Agricultural Land

The total amount of cultivable land in the twelve sand-covered areas has decreased significantly (Table 5). The villages lying very near to embankment have been totally damaged. These villages are Gitgram, Natungram, Bira and Brahman-Dihi where such characteristics are very common.

Table 5
Loss of Cultivable Land by Sand Splay

Mouja/village name	Total amount of cultivable land (ha)	Loss of cultivable land (ha)
Bhedia	335	61.66
Brahman-Dihi	164	21.66
Mallikpur	102	20.72
Natunhat	106	25.43
Nabagram	66.6	20.45
Bira	46	18.78
Bankai	44	10.64
Narega	75	22.38
Gitgram	110	30.80
Haripur	60	40.60
Rasulpur	110	40.20
Itanda/Natungram	90	40.20

(iii) Sedimentation and Filling Up of Water Bodies

The water bodies like tanks, marshy lands, canals have been completely filled up by sands and the situation has become so bad that it would not be possible to lift the sands in future. The area covered by water bodies in the individual villages has reduced and the productive occupations based on water bodies have destroyed. Therefore, this phenomenon is now looming large over the inhabitants as a poignant threat to their survival.

(iv) Other Problems

The cultivable lands are also being destroyed by the changing river courses in different parts of the lower basin. In some villages like Gitgram, Natungram, Rasulpur, Bangalpara, and Brahmandihi, the homestead areas have also been covered by sands. During summer, excessive heat generated from the sand splayed areas damages the crops of surrounding plots. It is also observed that the loose sands are transported away by the summer storms from the sand-covered areas and deposited over cultivable lands, thereby reducing soil fertility.

The flood affected people are now facing a lot of problems in sustaining their livelihood. The people of the above mentioned villages have become financially handicapped and they have no alternative measures to earn money for their subsistence. They are outmigrating from their own land in search of work to the other areas. The large landowners have been transformed to marginal farmers as a result of the loss of agricultural lands due to sand splay. During post-flood period there is no any materials simply like bamboos, straw and even soil in reconstructing their settlements.

Mitigation through Human Adjustment

Man exists in an essentially ecological relationship with his environment and has to live with a variety of natural hazards which threaten life and property. In the present situation, it is not possible to lift and transfer huge sand splay over agricultural land to other places. Therefore, some non-structural measures like change in landuse pattern, job opportunity schemes and relief insurance schemes should be introduced. Proper land management programme should be practised after considering sand depth, quality of soil nutrients, other physical properties and socio-economic demands. One important land management schemes, particularly for the sand covered area may be introduced according to sand depth and after few years the degraded lands may be transformed into suitable lands. The proposed plans are detailed below in Table 6.

Table 6
Proposed Plans for Mitigating the Problems in the Sand Splay Areas

Depth of Sand	Nature of Farming
6"-10"	Adding organic manure, Farm yard manure introduce root crops like potato, reddish, beet etc.
10"-18"	Cultivate kharif dhancha or green manure namely sunhemp, karang etc.
18"-24"	Suitable for agro-forestry by digging 18" diameter pit for multi purpose tree like subabul, akashmani etc. Surface to be covered by blackgram, greengram etc. after one year
More than 24"	Lifting of sand for construction and other purpose

On the other hand, to restrict sand splay hazards in other areas of the basin, some structural measures like protection of embankment and levees, channel enlargement, cutting of flood relief channels or diversion channels, project of flood storage reservoirs, etc. should be implemented. It can be concluded that man has to learn to live with the prevailing situation through proper adjustment, without tampering with natural hazards.

References

Dury, G.H. (1969), '*Hydrolic Geometry in Water, Earth and Man*,' edited by R.G. Chorely, Methun and Co., London.

Mukhopadhyay, Malay (1999), 'Flood Hazards in West Bengal: Mitigation through Human Adjustment,' published in *Flood Situation in West Bengal: Agenda for Action*, Vol. 1, edited by Directed Initiative, Calcutta.

Smith, K. and Tobin (1979), *Human Adjustment to Flood Hazards*, Longman, London.

Williams, Martin, A.J. et al. (1996), *Interactions of Desertification and Climate*, Arnold, London.

Word, R. (1978), *Floods: A Geographical Perspective*, Macmillan Press Ltd., London.

26

Impact of Human Activities on Land Degradation and Desertification in the Thar Desert: A Study Through Remote Sensing Technique

Mahesh K. Gaur and Hemlata Gaur

Thar Desert

The Thar Desert of western India is the most densely populated hot desert in the world. It lies between 24°40′ E to 30°12′ E longitudes, covering an area of 210,016 sq. km. which is about 64.1% of the Rajasthan state. It comprises of twelve districts of Rajasthan, namely, Barmer, Bikaner, Churu, Ganganagar, Hanumangarh, Jaisalmer, Jalor, Jhunjhunu, Jodhpur, Nagaur, Pali and Sikar. Typical characteristics of the Thar are: (i) slight off with astonishing variability; (ii) large diurnal variation of temperature and high maximum temperature during summer season causing great evapotranspiration; (iii) scanty xerophytic vegetation and total divestiture from agriculture making inhabitants animal dependent nomads and (iv) long and frequently occurring droughts and famines. Thar desert supports 1,75,09,490 persons. It is one of the most densely populated deserts of the world having a population density of 59 persons per sq. km.

Irrigation water is now available following the completion of the Indira Gandhi Canal project. However, dryland-farming practices, where water for plant growth is supplied entirely by rainfall, still predominates.

Desertification: Definition

Desertification is the impoverishment of terrestrial ecosystems under the impact of man. It is the process of deterioration in these ecosystems that can be measured by reduced productivity of desirable plants, undesirable alterations in the biomass and the diversity of the micro- and macro-fauna and flora, accelerated soil deterioration, and increased hazards for human occupancy.

Desertification is a land degradation problem of major importance in the arid regions of the world. Deterioration in soil and plant cover adversely affects nearly 50% of the land areas as the result of human mismanagement of cultivated and range lands. Overgrazing and woodcutting are responsible factors for most of the desertification of rangelands. Cultivation practices inducing accelerated water and wind erosion are most responsible in the rainfed croplands, and flood-irrigation leading to salinisation is the cause of the deterioration of irrigated lands. In addition to vegetation deterioration, erosion, and salinisation, land degradation and desertification effects can be seen in loss of soil fertility, soil compaction, and soil crusting. Urbanisation, mining, and recreation are having adverse effects on the land of the same kind as is witnessed on range, dry farming, and irrigated lands.

The 1977 Nairobi Conference drew attention to the desertification (UN Secretariat, 1977) as described below:

"...the diminution or destruction of the biological potential of the land, (which) can lead ultimately to desert-like conditions. It is an aspect of the widespread deterioration of ecosystems, and has diminished or destroyed the biological potential, i.e. plant and animal production, for multiple use purpose at a time when increased productivity is needed to support growing populations in quest of development. Important factors in contemporary society—the struggle for development and the effort to increase food production, and to adapt and apply modern technologies, set against a background of population growth and demographic changes— interlock in a network of cause and effect. Progress in development, planned population growth and improvements in all types of biological production and relevant technologies must therefore be integrated. The deterioration of productive ecosystems is an obvious and serious threat to human progress. In general, the quest for ever greater productivity has intensified exploitation and has carried disturbance by man into less productive and more fragile lands. Overexploitation gives rise to degradation of vegetation, soil and water, the three elements, which serve as the natural foundation for human existence. In exceptionally fragile ecosystems, such as those on the desert margins, the loss of biological productivity through the degradation of plant, animal, soil and water resources can easily become irreversible, and permanently reduce their capacity to support human life. Desertification is a self-accelerating process, feeding on itself, and as it advances, rehabilitation costs rise exponentially. Action to combat desertification is required urgently before the costs of rehabilitation rise beyond practical possibility or before the opportunity to act is lost forever" (UN Conference to Combat Desertification, 1977).

A definition of desertification should recognise that it is a land degradation process that involves a continuum of change, from slight to very severe degradation of the plant and soil resource, and is due to man's activities. Dregne (1986) discusses three key desertification processes in arid lands:

- deterioration of vegetative cover due to overgrazing, wood cutting, and burning;
- wind and water erosion resulting from improper land management; and
- salinisation due to improper use of irrigation water.

Based on above definitions, desertification could be characterised by the following components: *climate*: increasing aridity (diminishing water/moisture supply); *hydrological processes*: runoff becoming more irregular or drying-up; *morphodynamic process*: intensification of distinct geomorphological process (accelerated soil erosion by wind and water); *soil dynamics*:

desiccation of soils and accumulation of salt; *vegetation dynamics*: decline of vegetative cover or biomass.

Methodology

Land is the major non-renewable resource and faces the biggest threat of degradation. Land resources of the Thar desert are degrading at an alarming rate and causing environmental and social problems and concerns. The loss of valuable soil due to water and wind erosion has been a matter of serious concern for the inhabitants of the Thar. Wind erosion is quite active in this region, thereby threatening the very agricultural sustainability and the environment in the context of desertification.

Sustained activities on the low fertility land of the Thar desert have impaired the productivity at many a places, causing land degradation. Due to cultivation on marginal and sub-marginal lands, vast areas have been categorised as poor due to low organic matter and hence poor nutrient supply capacity. On the other hand, the abuse of community forest resources (*orans* and *gauchars*) and rangelands has caused serious disturbances in the natural ecosystems of the Thar and is leading towards loss of biodiversity and fertile top soil.

Present study on the desertification and land degradation under the desert ecosystem has been using the Survey of India topographic sheets and IRS-1B and 1D imageries of 1:250,000 and 1:50,000 (False Color Composites—FCCs) and extensive ground truthing. District census handbooks and socio-economic data published annually by the state government were also extremely helpful. Based on those observations, the boundaries of different micro-ecosystems were delineated for finding impact of human activities. The morphological changes in these micro-climate ecosystems resulted due to various degradational processes like aeolian and fluvial erosion/deposition, salinity, floods, waterlogging, cultivation on marginal and sub-marginal lands and sand dunes, over-exploitation of resources, burgeoning livestock population, tractorisation of cultivation, and sporadically variable rainfall. The vertical and horizontal extent of the hazards created by the accelerated desertification and land degradation processes in these ecosystems under study have also been measured.

Results and Discussion

There is an acute need for systematic analysis and geo-referencing of the various forms of multiple landuse in the Thar desert. Results obtained from the interpretation and analysis of the satellite and secondary data supplemented with ground truthing indicate conditions of present scenario. Unfavorable climatic conditions of the Thar, allow rainfed farming except where canal water has reached (Bikaner, Hanumangarh and Jaisalmer). Due to frequently recurring droughts and famines, income from agriculture is low, even not assured. Often it is a loss-making enterprise. Consequently, unemployment in rural areas has forced people to shift from villages to cities and to changing their traditional occupation. A variety of human induced pressures, as identified from the IRS satellite FCCs and ground truthing, have caused the following major land and environmental problems:

- Vanishing of community lands (*orans* and *gauchars*).
- Soil erosion.
- Surface crusting and soil compaction.
- Loss of biodiversity.
- Poor quality livestock production.
- Waterlogging and salinisation.
- Mine spoilt lands.

Water Resources

The white and dull grayish-white patches, of the salinised areas, on the FCC are distinct and easily identifiable. A major portion of the Thar desert (125,008 sq kms) has saline groundwater. About 5% to 8% area has high-fluoride content in the groundwater. Around 65% of the groundwater is already exploited for drinking and agricultural uses. Hence, places with the groundwater level of 30 to 80 feet are now unable to exploit groundwater until they reach the underground depth of 300 to 500 feet.

Tubewell and Commercial Agriculture

About fifteen years ago, Mathania and Tinwari areas of Jodhpur district boasted of wells with a water level of merely 5 to 60 feet. Indiscriminate drilling of tubewells by the rich farmers in the last two decades have caused the drying-up of the open wells, belonging to the poor, as well as the community wells. Currently, a single rich farmer owns many tubewells that exploit the groundwater at its maximum by irrigating crops like chilli (*Capsicum annum*) that needs watering 30 to 40 times during a normal crop. It is pertinent to illustrate here that about 100 thousand tubewells have been working in Jodhpur district alone, of which about 40,000 have been dug during last one decade. This area has been classified under dark zone.

Water Pollution

Rivers of Thar are polluted because of the faulty industrial policies. Luni, the longest river of the Thar, and other perennial rivers like Jojri and Bandi are highly polluted due to the industrial effluents. They have polluted all the wells situated on their coasts or connected through their channel. About 5,000 open wells and over 100,000 acres of land are polluted and have already been classified as wasteland. Two decades ago, a large number of families used to cultivate wheat, mustard, *jeera* (*Cuminum cyminum*), and different kinds of vegetables there. However, the harvest in this entire region has reduced to merely 18 to 20% because of pollution.

The only grass that grows here is known as 'Dabh' that is not even eaten by the animals. Similar problem of pollution exists throughout Jodhpur-Salawas, Balotra-Sindhar, Pali-Jetpura regions. Water pollution has drastically eroded the natural resource base of Thar. About more than 20 million gallons of water is being polluted everyday in Jodhpur, Pali, and Balotra areas.

Destruction of Catchments and Health Hazards Due to Stone Mining and Health Hazards

Unregulated mining has adversely affected the land and water based ecology of Thar because entire catchment area of rainwater harvesting and the auxiliary network of wells, *baoris, jhalaras, nadis*, ponds, lakes and canals has been totally destroyed and top fertile soil is also eroded due to continuous stone mining operations.

Land Resource

Vast brownish-red to brown areas on the satellite imageries indicated the horizontal expansion of sand dunes in the Thar desert. Such sand dune soils of the Thar have certain unique features:

- Lack of organic matter.
- Normal to very high level of potash.
- Over-exploitation of the land.
- Less than normal level of nitrogen as well as phosphorous.
- Weak water-retention capacity.
- An abundance of hard pan at varying levels under the ground, covering an extremely large area, which checks further percolation of water.
- An extremely large saline land area.
- Around 30 to 40% land area is full of sand dunes.
- Fast blowing winds (with the maximum wind velocity of 110 km/hr) causing acute soil erosion problem.

The continuous exploitation and misuse of land area has severely diminished its fertility. It has been calculated that one bigha land (one bigha = 132 x 132 feet), which used to yield four quintals of barley, is unable to produce even one sack of barley today.

Excessive Farming

Ploughing in Thar desert is full of difficulties. Around 28.61% land was used for cultivation in 1951, which increased to 45.5% by 1971, and 68.3% by 1991. The interpretation of census data has revealed that 11% households owned 50% of the cultivable land, whereas 47% households owned a mere 10%. Over-use of fallow land or its exploitation has diminished its fertility and gradually the land has almost turned into a wasteland.

Soil Erosion-Tractor Cultivation

Soil erosion has become a grave problem since the last three to four decades. Ploughing with the help of tractors is one of the main causes for land degradation. There were 112,175 tractors available in the Thar region in 1997, which is 51% of total tractors available with the farmers in the entire state of Rajasthan.

Three decades ago, animals like bullocks, camel, donkey, mule, etc. were utilised in ploughing. Previously, there used to be several trees and bushes in the fields (farms) and the farmers used to make vegetative barriers by putting bushes at a distance of 30 to 40 m (against the wind direction) to check soil erosion. However, the introduction of tractors, especially the usage of disc-ploughing technology has de-rooted all the on-farm trees and bushes. Even the fencing has been stopped as it proved a hindrance in tractor ploughing. The lack of rootstock resulted in soil-erosion, as there is nothing to hold the topsoil in place.

Tractor ploughing has in fact become a bane for the soil of Thar. Villagers have experienced that the tractor ploughing has destroyed a grass called *Sevan* (*Lasirus*) that is not only a nutritious fodder but also very effective in holding the soil. Because of its use, irreparable devastation has occurred in the Thar.

Flood-Irrigation

A mega-project—called Indira Gandhi Canal Project (IGCP)—has come up to irrigate the vast areas of Thar desert. Fifteen million hectares of cultivable land (including proposed areas) has been covered by this project in Ganganagar, Hanumangarh, Churu, Bikaner, and Jaisalmer districts. Canal irrigation mainly comprises flooding the farmland through irrigation channels. The availability of water from the IG canal has boosted the agricultural production manifold. This kind of irrigation is mainly used for high incentive commercial crops like mustard, cotton, groundnut, sugarcane, etc.

that require excess of water. Farmers have started growing crops of rice also in this region, which is very harmful for the desert biodiversity as well as for the land capability.

The desert soils are prone to salinity and have hard rock underneath. The canal irrigation has been going on for more than 70 years in Ganganagar. This has converted large cultivable areas into marshy lands because of rise in water table. About 20% of the total area in the Rawatsar tehsil of Hanumangarh and 70% of the total area in Ganganagar is waterlogged and marshy (1998 and 1999) (Tables 1 and 2). This is wreaking havoc in these districts because several villages have practically lost their entire cultivable land.

Soil of Thar desert does not have the capacity to cope with the flood irrigation. Thinking in the long-term, it is fatal to allow cultivation of crops which over-use water and use an irrigation system that floods the farm.

Table 1

District-wise Area under Degraded Lands (1996-97)

S. No.	District	Area	Percentage
1.	Barmer	3,71,201	13.1
2.	Bikaner	8,13,619	31.3
3.	Churu	14,725	0.87
4.	Ganganagar	52,927	7.0
5.	Hanumangarh	9,827	0.990
6.	Jaisalmer	31,58,181	82.2
7.	Jalor	1,13,167	10.6
8.	Jhunjhunu	21,668	3.7
9.	Jodhpur	1,73,879	7.6
10.	Nagaur	73,026	4.1
11.	Pali	1,92,613	15.5
12.	Sikar	26,488	3.4
	Total area	50,59,509	100.0

Table 2
Land Degraded Due to Human Activity in the Ganganagar District (1998)

	Area (ha)	% of district area	% of total wastelands
Waterlogged	12498.42	1.14	7.10
Land affected with salinity	22.53	0.002	0.01
Sandy area	72294.72	6.59%	41.06
Total Land area	176072.51	100	100

Land Degarded Due to Human Activity in the Rawatsar Tehsil of Hanumangarh District (1997)

Waterlogged area	1.62%
Marshy land	0.33%
Mining/industrial waste	0.95%

Table 3
Grazing Lands Area in Thar Desert (1996) (in hectares)

S. No.	District	1991	1996
1.	Barmer	205561	205010
2.	Bikaner	48975	47041
3.	Churu	46749	46583
4.	Ganganagar	2830	2894
5.	Hanumangarh	8811	8844
6.	Jaisalmer	107593	107428
7.	Jalor	46106	45734
8.	Jhunjhunu	40539	40433
9.	Jodhpur	124602	124306
10.	Nagaur	71796	71613
11.	Pali	90159	90060
12.	Sikar	43351	41656
	Total Area	910002	832902

Source: IRS-IB/ID Satellite data.

The Growing Desert

Around 30 to 40% of the entire Thar desert areas are nothing but sand dunes. Barmer district has not only the maximum sand dunes but the biggest as well. Since last two decades, these dunes are losing their native natural vegetative cover at a rapid pace. Depleting vegetation and expansion in agricultural activity are responsible for this development.

Devoid of natural vegetative cover, such land is loosing more topsoil every year. Hence the formations of *in situ* sand dunes, which is adversely affecting the crops. These barren (uncultivable) dunes are called *Thathron* in the local dialect.

Plant Resource

The yellowish to grayish-yellow tint covering vast areas indicates an area without any vegetative cover or areas with scrub. Thar desert has scarce vegetation in terms of grass, fodder, firewood, and agricultural produce.

Around five decades ago, only 28.61% land was under cultivation and that too with the sole purpose of keeping half of the cultivable land fallow for 3 to 5 years. The remaining land was used for growing forests, pastures, and utilised for catchment areas of *nadis*. Official data reveals that only 5% of the total land was used for agriculture whereas the remaining 95% was used as community land wherein grass bushes and trees were grown. The Barmer district has earmarked 14.6% land for agricultural purposes and 42% community land for development of vegetation.

Loss of Common Property Resources

Common property resources contribute in a variety of ways to the village economy. They provide local communities with fuel, fodder, water, and food for consumption within the home and for sale in the market place. Local communities no longer have shared responsibilities for the management and conservation of forests and pasture lands.

Pastures: Pastures are part of an age-old tradition of Thar desert (Table 3). A special piece of land was attributed for grazing and fodder production in almost all villages. *Sewan, Dhaman, Ghadhia, Bhurat, Bekaria*, are some of the major grasses of this region that require very little rain for their growth. But, the development policies have neglected this age-old tradition and hence the fodder production has gone down. Jaisalmer and Barmer till recently boasted of large land areas where *sewan* grass was grown but such areas are fast depleting due to introduction of canal irrigation. The rate of depletion during 1987 was quite imaginable when Thar desert witnessed its worst drought of this century. *Sewan* was found in abundance in Jaisalmer.

Village Forests (Orans): Orans—the traditional village forests have been heavily neglected in the Thar desert. In 1988, 23207 hectares of land belonging to 85 villages of Jodhpur-Barmer-Jaisalmer districts comprised 586 hectares (2.5%) of good quality forests. Around 12.5% land or 2953 hectares consisted of partial forests and the remaining 85.5% land was totally devoid of vegetation and hence barren. It was also revealed that all kinds of natural xerophytic shrubs—*Fog, Khejari* and *Rohida*—have drastically reduced. This study revealed a very sad picture of *orans* in the Thar desert. Common shrubs like *Fog* (*Calligonoum polygonides*), *Khejari* (*Prosopis cineria*), *Rohida* (*Tecomella undulata*), and grasses like *Bhurat, Lampara, Bekaria, Diabatti, Motha,* and *Dhania*; and perennial grasses like *Tania, Dhaman, Khargos-chunti, Chirio-gass, Tantia-gass, Dab, Sonali, Kali* and *Safed Bui* (*Aerva tomentosa*), *Dhamasio, Kali-murgi,* and *Moda-Dhaman* were on the verge of extinction. Similarly, a number of shrub and tree species were also threatened. *Vilayati Babool* (*P. juliflora*) can be seen everywhere. The major reasons behind this depletion, even cited by the villagers, have been the ever-growing population (of man and livestock), monetary outlook, and the sheer neglect of the traditional village forests by village communities as well as in the development policies.

Livestock Resource

The livestock of Thar amounted to 9.4 million in 1951, 15.5 million in 1971, which gradually increased to a total of 28.6 million by 1997. In the year 1951, the sheep and goat population of Thar formed 57% of the entire animal population, which shot up to 70% by the year 1972. It stagnated at the same percentage in quinquennial livestock census, 1997.

Conclusion

Sound management of the environment, from the global level to sub-regional, requires good assessment of situations. But, in order to produce good assessment, it is necessary to have timely and reliable data and information. The kind of data and information needed for a particular decision depends upon the decision itself. Remote sensing technique allows an efficient flow of data and information from the data generation to the decision-making level.

There is both competition and complementarity of landuses in a spatial context. Traditional land users in Thar desert have often remained versatile and adapted to the local rainfall variability through year-to-year changes in agricultural practices and herding. Cultivation, livestock rearing and wood-gathering practices as well as the use of the technology have all been major causes of and contributors to, the desertification and land degradation processes in the Thar desert. Human activities have influenced the surface by: (a) reducing vegetation cover (by overgrazing, cultivation, deforestation), (b) decreasing the roughness thereby increasing wind speeds, (c) altering soil moisture patterns, and (d) burning vegetation and dislodging dust at the surface.

Acknowledgement

The author first expresses his gratitude to the University Grants Commission, Ministry of Human Resources Development, Government of India for providing financial support through the award of research project. This research paper is the outcome of a number of studies conducted by the author. Thanks are also due to the EMSD, Jodhpur for providing facilities to use their infrastructure.

References

AAAS. 1970. Pub. No. 90. Washington.

Agrawal, S.K. et al. 1996. *Biodiversity and Environment*. APH, New Delhi.

Bharara, L.P. 1999. *Man in Desert*. Scientific, Jodhpur.

Dregne, H.E. 1997. Desertification: Man's Abuse of Land. *Journal of Soil and Water Conservation*. 33:11-14.

Dregne, H.E. 1982. Desertification in the Americas. 'Desertification and Soil Policy' Trans. 12th International Congress of Soils, New Delhi.

CAZRI. 1994. Mimeograph.

Dhir, R.P. and Gaur, M.K. 1998. Landuse/Landcover Study of Nohar, Rawatsar and Bhadra tehsil, Hanumangarh district. Submitted to National Remote Sensing Agency (NRSA), Department of Space, Hyderabad.

Dhir, R.P. and Gaur, M.K. 1998. Resource Assessment Plan of Nohar, Rawatsar and Bhadra Tehsil, Hanumangarh District. Submitted to NRSA, Department of Space, Hyderabad.

Dhir, R.P. and Gaur, M.K. 1998. Land Resource Development Plan of Nohar, Rawatsar and Bhadra Tehsil, Hanumangarh district. Submitted to NRSA, Department of Space, Hyderabad.

Dhir, R.P. and Gaur, M.K. 1998. Water Resource of Nohar, Rawatsar and Bhadra Tehsil, Hanumangarh district. Submitted to NRSA, DOS, Hyderabad.

Gaur, M.K. 1992. *Impact of Drought: A Biogeographical Study*. University of Jodhpur, Jodhpur.

Gaur, M.K. 1992. Impact of Drought on Desertification Process in Northwest Rajasthan Desert (based on remote sensing technique). Unpublished Ph.D. thesis. JNV University, Jodhpur.

Gaur, M.K. 1992. 'Economic Losses Due to Drought in Jodhpur District' in *Environment Conservation, Management and Planning*. Chug Publications, Allahabad.

Gaur, M.K. 1992. 'Environmental Degradation Due to Over-exploitative Human Reactions During Drought' in *Environment Conservation, Management and Planning*. Chug Publications, Allahabad.

Gaur, M.K. 1994. 'Shadows of 1987 Drought: A Catastrophe of Thar Desert' in *Environmental Degradation and Global Awakening*. Chug Publications, Allahabad.

Gaur, M.K. 1995. 'Environmental Crisis in the Thar Desert: Population, Pollution and Resources' in *Environmental Conservation for Global Development*. Chug Publications, Allahabad.

Gaur, M.K. 1996. Bio-Geo-Environmental Status Report of Textile Industrial City of Pali. Submitted to IIEE, New Delhi.

Gaur, M.K. 1997. Biodiversity Assessment of Thar Desert (a case study). WWF-India and SDS, Jodhpur.

Gaur, M.K. and Gaur, Hemlata. 1997. Environmental Survey of Thar Desert (A case study of Bhalu watershed). Submitted to GVVS, Jodhpur and Wells for India, U.K.

Gaur, M.K. and Gaur, Hemlata. 1999. Wasteland Mapping of Ganganagar District. Submitted to NRSA, DOS, Hyderabad.

Gaur, M.K. 1999. Appraisal of Biotic and Abiotic Resources in an Arid Environment through the Application of Remote Sensing Technique. Department of Geography, JNV University, Jodhpur (with B.B.L. Sharma).

Gaur, M.K., and Gaur, Hemlata and Sharma, B.B.L. 1999. Mapping of Natural Resource Bases in an Arid Environment through Carto-techniques. ISRS National Conference, Goa.

Gaur, M.K., and Gaur, Hemlata and Sharma, B.B.L. 1999. Man and Natural Resource Bases Relationship in an Arid Environment (A Case Study Through Remote Sensing Technique). UGC National Seminar at Churu.

Gaur, M.K., and Gaur, Hemlata and Sharma, B.B.L. 2000. Urbanization and the Environment (A case study of Rajasthan state). International Conference on Asian Urbanization, Chennai.

Gaur, M.K. and Gaur, Hemlata and Sharma, B.B.L. 2000. Sustainable Landuse Planning of Arid Environment through Satellite Remote Sensing. 21st Indian Geographical Congress of NAGI at Nagpur.

Gaur, M.K. 2000. Remote Sensing in Monitoring Ecological Aspect of Drought and Desertification Process in North-west Rajasthan desert. IGC of International Geographical Union at Yerevan.

IEMSD. 2000. Status paper "The Thar Desert Ecosystem: Some Development Related Issues". Institute of Environment Management and Sustainable Development, Jodhpur.

NAEB. 1997. Sacred Groves of Rajasthan. Ministry of Environment and Forests, Government of India.

Sharma, H.S. and Sinha, A.K. 1990. *Ecology of Land and Water Management*. Kuldeep, Ajmer.

Sinha, R.K. et al. 2000. *Desert Management and Desertification Control*. INA Shree, Jaipur.

Thukral, R.K. et al. 2000. *Desert Management and Desertification Control*. INA Shree, Jaipur.

Thukral, R.K. 1998. *Rajasthan at a Glance*. Jagran, Kanpur.

UNCOD. 1977. India Country Report.

List of Contributors

A.K. Kulkarni, Indian Institute of Tropical Meteorology, Dr. Homibhabha Road, Pasan, Pune, Maharashtra, INDIA.

Ananya Taraphder, Department of Geography, University of Burdwan, Burdwan, West Bengal, INDIA.

Annegret Haase, Centre for Environmental Research, Department of Applied Landscape Ecology, UFZ-Umweltforschungszentrum, Leipzig-Halle GmbH, PF2, 04301 Leipzig, GERMANY.

Avijit Gupta, School of Geography, University of Leeds, Leeds LS2 9JT, UNITED KINGDOM.

B.N. Mandal, Indian Institute of Tropical Meteorology, Dr. Homibhabha Road, Pasan, Pune, Maharashtra, INDIA.

D. Das, Department of Environmental Science, University of Kalyani, Kalyani, West Bengal, INDIA.

Dagmar Haase, Centre for Environmental Research, Department of Applied Landscape Ecology, UFZ-Umweltforschungszentrum, Leipzig-Halle GmbH, PF2, 04301 Leipzig, GERMANY.

Debasish Sarkar, Department of Economics, Hooghly Mohsin College, Chinsura, West Bengal, INDIA.

Gear M. Kajoba, Department of Geography, University of Zambia, PO Box 32379, Lusaka, ZAMBIA.

Hemlata Gaur, Department of Geography, S.B.K. Government P.G. College, Jaisalmer, Rajasthan, INDIA.

K. Gupta, Department of Geography, Visva-Bharati University, Santiniketan, West Bengal, INDIA.

Kuntala Lahiri-Dutt, Department of Geography, University of Burdwan, Burdwan, West Bengal, INDIA.

Lothar Linde, Centre for Environmental Research, Department of Applied Landscape Ecology, UFZ-Umweltforschungszentrum, Leipzig-Halle GmbH, PF2, 04301 Leipzig, GERMANY.

Mahesh K. Gaur, Department of Geography, S.B.K. Government P.G. College, Jaisalmer, Rajasthan, INDIA.

Malay Mukhopadhyay, Department of Geography, Visva-Bharati University, Santiniketan, West Bengal, INDIA.

Mallanath Mukherjee, Department of Geography, Visva-Bharati University, Santiniketan, West Bengal, INDIA.

Manjari Sarkar(Basu), Department of Geography, University of Burdwan, Burdwan, West Bengal, INDIA.

Maria Sala, Department of Physical Geography, Faculty of Geography and History, University of Barcelona, 08028 Barcelona, SPAIN.

Mesbah-us-Saleheen, Department of Geography and Environment, Jahangirnagar University, Savar, Dhaka 1342, BANGLADESH.

Mirza Mafizuddin, Department of Geography and Environment, Jahangirnagar University, Savar, Dhaka 1342, BANGLADESH.

Motilal Ghimire, Central Department of Geography, Tribhuban University, Kathmandu, NEPAL.

N. Batnasan, Institute of Geography, Mongolian Academy of Sciences, Ulaanbaatar-210620, MONGOLIA.

N.K. De, Department of Geography, University of Burdwan, Burdwan, West Bengal, INDIA.

Nageshwar Prasad, Department of Geography, University of Burdwan, Burdwan, West Bengal, INDIA.

Narendra Raj Khanal, Central Department of Geography, Tribhuban University, Kathmandu, NEPAL.

Oliver Spott, Centre for Environmental Research, Department of Applied Landscape Ecology, UFZ-Umweltforschungszentrum, Leipzig-Halle GmbH, PF2, 04301 Leipzig, GERMANY.

Onkar Prasad, Department of Geography, K.N. Government College, Gyanpur, Uttar Pradesh, INDIA.

Prasanta K. Jana, Department of Geography, University of Burdwan, Burdwan, West Bengal, INDIA.

R.B. Sangam, Indian Institute of Tropical Meteorology, Dr. Homibhabha Road, PASAN, Pune, Maharashtra, INDIA.

R.G. Patil, Department of Geography, SNDT Women's University, Pune, Maharashtra, INDIA.

R.Y. Singh, Department of Geography, University of Zambia, P.O. Box 32379, Lusaka, ZAMBIA.

S. Gangopadhyay, Anthropological Survey of India, 2, Ripon Street, Kolkata, West Bengal, INDIA.

S.N. Chatterjee, Palli Charcha Kendra, Sriniketan, West Bengal, INDIA.

Saswati Kapat, Department of Geography, Visva-Bharati University, Santiniketan, West Bengal, INDIA.

Subrata Ghosh, Department of Geology, Government College, Durgapur, West Bengal, INDIA.

Susmita Ghosh, Qr. No. B-9, Government College Campus, Durgapur, West Bengal, INDIA.

Sutapa Mukhopadhyay, Department of Geography, Visva-Bharati University, Santiniketan, West Bengal, INDIA.

Thilo Weichel, Centre for Environmental Research, Department of Applied Landscape Ecology, UFZ-Umweltforschungszentrum, Leipzig-Halle GmbH, PF2, 04301 Leipzig, GERMANY.

V.C. Jha, Department of Geography, Visva-Bharati University, Santiniketan, West Bengal, INDIA.